Geographies, Mobilities, and Rhythms over the Life-Course

By thinking in terms of the geographies of mobilities, we are better able to understand the central importance of movements, rhythms, and shifting emplacements over the life-course. This innovative book represents research from a new and flourishing multidisciplinary field that includes, among other things, studies on smart cities, infrastructures, and networks; mobile technologies for automated highways or locative media; mobility justice; and rights to stay, enter, or reside. These activities, cadences, and changing attachments to place have profound effects—first on how we conduct or govern ourselves and each other via many social institutions, and second on how we constitute the spaces in and through which our lives are experienced. This scholarship also has clear connections to numerous aspects of social and spatial policy and planning.

Elaine Stratford lives in Hobart, where she is a member of the Discipline of Geography and Spatial Sciences at the University of Tasmania.

Routledge Advances in Geography

Geographies, Mobilities, and Rhythms over the Life-Course

Adventures in the Interval

Elaine Stratford

NEW YORK AND LONDON

First published 2015
by Routledge
711 Third Avenue, New York, NY 10017

and by Routledge
2 Park Square, Milton Park, Abingdon, Oxon OX14 4RN

*Routledge is an imprint of the Taylor & Francis Group,
an informa business*

© 2015 Taylor & Francis

Library of Congress Cataloging-in-Publication Data
CIP data for this book has been applied for.

ISBN: 978-0-415-65936-9 (hbk)
ISBN: 978-0-203-07497-8 (ebk)

Typeset in Sabon
by Apex CoVantage, LLC

To Lewis and Michael

"What is, is a refrain. A scoring over a world's repetitions. A scratching on the surface of rhythms, sensory habits, gathering materialities, intervals, and durations." (Kathleen Stewart 2010, 353)

Contents

Figures

Acknowledgments

The pages that follow represent a capstone—but not, I hope, a terminal point. Doubtless, I have been circling around this book for some time. Reflecting on its conception, I think it must have been in my head for more than two decades. Yet, it is only in the last five years that I have felt moved to give it form, collecting materials and pondering their consequences, and then writing sketches that would later become chapters. Part of the reason for the delay is that in those years I have come (back) to geography in an emotional sense, at the same time re-cognizing both its particular disciplinary strength and aesthetic, and its generosity—for the present work is markedly interdisciplinary in scope, a choice both liberating and perilous if read ungenerously . . . and generosity is a hallmark theme in the pages that follow. Either way, the work is now done, and I hope that it makes a small and valuable contribution to a much wider and profoundly important set of questions about the ways in which we live, encounter ourselves, each other, and the more-than-human world.

Without doubt, my experience of writing has been enriched by the support of a number of people. I am most grateful to Max Novick, Christina Faria, and the team at Routledge for enabling me to make this contribution to the Advances in Geography series, and for the timely, professional, and caring manner in which they have engaged with me. In addition, I want to acknowledge the authors on whom I draw in this work; they do not know it yet (and some never may), but my scholarship has been deeply enriched by engaging with them. I am deeply grateful to Mahni Dugan, who has been with me since the start of the project. While writing her own immensely powerful doctoral work, Mahni provided careful and considered research assistance to me, as well as eagle-eyed copy-editing, and sensitive engagement with libraries and those who have provided images for the text. Those images have been carefully prepared for publication by Nicola Stratford, who explained with good humor why I should not be allowed near such labors unsupervised, and who is among the most generous of souls. I am appreciative of the work done for me by staff in the Morris Miller Library at the University of Tasmania by obtaining a range of documents, and especially to Christine Evans for clear and patient guidance with Endnote. For timely and constructive feedback on particular chapters or ideas, I am most grateful to Susan Dodds, Andrew Wells, Richard Coleman, Elizabeth Jones, Danielle Drozdzewski, Ian Buchanan Gordon Waitt, and Jeff Malpas. Caroline Cottrell has read and reread drafts as they appeared in her inbox at all hours, and has provided invaluable and well-placed critique along the way. Many years ago, she showed the same generosity with a certain doctoral thesis, as I recall—thank you. I want, in addition, to acknowledge Godfrey Baldacchino and Elizabeth McMahon. Although, as I write this, neither has yet seen the manuscript, Godfrey and Liz have been close collaborators on projects in recent years, and both have significant track records in publishing; they may not yet be aware of this fact, but they helped me find my voice again. Notwithstanding the ways in which the work has been enriched by the input of all of these good people, any shortcomings are, of course, my own.

During much of the period when the book has been in preparation, I was the Head of School, Geography and Environmental Studies at the University of Tasmania and, over the course of two years to the end of 2013, we prepared for a restructure that has involved new affiliations and new opportunities. To my colleagues and especially to Stewart Williams and Aidan Davison, fellow travelers in human geography, my sincerest thanks for supporting me to pursue this endeavor while I was leading that restructure process; I enjoy being back as one of the crew.

Through the pages of the book will also be found reference to the observation Aristotle and subsequent others make about the importance of friends in our lives. I have been blessed in this respect by knowing that several people would be there if I needed them, so thank you Dianne, Jayne, Jill, Siobhan, Krissy, Claudette, Denbeigh, Tad and Joacia, Kerry, Julie, and Nicki. I am much indebted to David, Anjeleca, Jeff, Maree, and Steve for keeping me aligned. Finally, there is my home front, wherein my deepest friendships reside. There, one will find Megsie and Monty, two dogs who are particularly good listeners, and keep me grounded (literally, for woe betide me if a daily walk is not factored into writing). Most importantly, there are three menfolk—my partner, Philip, and sons, Lewis and Michael—without whom everything would make much less sense. Thank you for always supporting me in my adventures, for tolerating my odd rhythms, my intense desire always to be on the move, and my clear passion for the subject matter of geography.

1   Adventures in the Interval

On my iPad is a folder that I have labeled Be, in which are various apps—tools entangled in progressively more mobile lives which may seem antiquated all too soon; nonetheless, presently they inform my practices as a geographer. The apps are symptomatic of the digitization of my world. Using bits of software to transport gigabytes of digitized work and using an international adaptor, keyboard, and iPhone, I need not carry a weighty laptop nor piles of tatterdemalion paper, as I once did. The technologies I have adopted also influence how I play as well, reading books on Kindle while listening to music on iTunes, dipping in and out of apps that keep my brain fit, track my daily walking, or test my patience with Sudoku.

These tools give effect to the variegated geographies, mobilities, and rhythms of my working day and I feel fortunate that their power to enable connection also enhances the attachment I feel to several places—home, neighborhood, work, and sites around the globe where people I care about are located. Yet, as Edward Casey (1996, 26, original emphasis) has noted in this regard, "place is more an *event* than a *thing* to be assimilated into known categories". And all that brings me back to the Be folder. Inside it are several diversions, including a delightful arts project called the *0 to 100 Project*, whose provenance and purposes are explained on the Internet. Its sponsors are two Canadian companies—Flash Reproductions and Up Inc.— whose staff work together on electronic bookbinding techniques. As part of one collaboration, in 2010 they commissioned the Canadian photographer and filmmaker Sandy Nicholson to work with them. Over the equivalent of ten days of photo shoots, Nicholson worked in Sydney and Toronto with three hundred people. One hundred and one of his images of some of them have been captured in an exhibition, book, YouTube clip, website, and app [http://0to100project.com/]. Accompanying the images are quotes from the participants, sharing their feelings about their age and stage in the life-course.

Initial exposure to the *0 to 100 Project* App is a rapid-fire nine-second slide show of headshots in front of a uniform background, starting with a howling infant boy named Robin Walker and ending with a smiling centenarian named Marguerite Davis (Nicholson 2011). Robin, being under the age of one year, says nothing, but his face, crumpled in a cry, speaks volumes: it is as if he has the cares of the world upon him. On the other hand, Marguerite's remark to Nicholson was that "You don't have to worry about anything. You let other people do the worrying".[1] It appears that she had found a place of contemplative peace and come to flourish. Between the two are images of fifty-five females and forty-four males, one person for each age between one and ninety-nine, each with telling thoughts to share.

In commenting on his work, Nicholson (2011, n.p.) has reflected on how different the outcome would have been had the photographs "been shot in different countries, such as Afghanistan or Malawi". Assuredly, life expectancy is forty-four in the former nation and fifty-six in the latter, and it is fair to say that the project would have had different outcomes in Australia and Canada had other participants been involved. For example, life expectancy is sixty-seven for Indigenous Australian men and seventy-two for Indigenous Australian women. On average, that is ten years younger than for the rest of the population (Australian Bureau of Statistics 2010). Statistics Canada (2010b) records life expectancy there at seventy-nine for the total male population and eighty-three for the total female population. Projected life expectancy among Canadian Aboriginals is lower. By 2017, it is expected to be sixty-four years for Inuit men and seventy-three for Inuit women; seventy-four for First Nations men and seventy-eight for First Nations women. Across the board, that is an average increase of only a year or two on figures for 2001 (Statistics Canada 2010a).[2] Such matters starkly show the relative and scalar effects of the social determinants of health inequalities. In this respect, Michael Marmot (2005) suggests that it should *not* be a certainty that life expectancy varies *across* countries by forty-eight years, and *within* countries by twenty years.

In my reading, the *0 to 100 Project* is a sensitive and imaginative interpretation of innumerable possibilities across the intervals that comprise the life-course. So, for example, accounting for the demographic privilege that is implied by the infant Robin Walker's Australian or Canadian identity, should he sit down to his hundredth birthday dinner in 2110, his life may well have been experienced as a series of relatively stable, if not durable, emplacements punctuated by stints of travel both short and long distance, and short and long term. Woven through those experiences may be the most astonishing geographies, mobilities, and rhythms. Robin's conception involved these; so, too, his gestation and birth; his education, play, work, senescence, and death will be entangled in others. Between these two points are countless changes in life—across generations, situations, and what British writer Ford Madox Ford (1905, 59) called "spacious times".

The essays in this collection are critically concerned with that variegated path—with our journeys over the life-course, their geographies, mobilities, and rhythms, and the extent to which their conduct enables us to flourish. In the balance of the first chapter, my goal is to explain and justify this agenda.

MARKING THE TERRAIN

Mobilizing Certain Invitations

I intend to take up a call "for a more finely developed politics of mobility" that attends such things as "motive force, speed, rhythm, route, experience, and friction" (Cresswell 2010b, 17). My method of approach will be to weave together my own research and experiences, and the work of others in a capstone manuscript bringing together fifteen years of thinking on particular problems. Allied to this task, I mean to mobilize contingent responses to three other invitations. One asks researchers to engage more often and more fully with emotional geographies (Anderson and Smith 2001). A second invites us to consider the significant worth of auto-ethnographic approaches to research (Ellis, Adams, and Bochner 2011; Ellis and Bochner 2000). The third and most recent invitation was made by Tim Edensor (2014, 169), who observed that "studies of the rhythms and other temporalities of mobility are in their infancy . . . Future study could explore in greater detail these multiple and contested mobile rhythms and the values and practice surrounding them".

I hope to add to *and connect* such works by considering two questions. The first is about the 'conduct of conduct', which Michel Foucault conceived in terms of the subject, power, behavior [*se conduire, la conduite*], and leading [*conduire*] (see the translator's note to Foucault 1982, 789). The second concerns what it means to flourish, a state termed *eudaimonia* by Aristotle (350 BCE(b); see also Ghaye 2010). My point of departure is to ask *as we move through our lives, how do we conduct or govern ourselves and each other, and in doing so how are we to constitute the spaces and flows for a life that is flourishing?* I ask this question because the geographies, mobilities, and rhythms of our lives invite concerted reflection. For example, consider work on the ethics of mobility by Sigurd Bergmann and Tore Sager (2008, 1), in which they note that the "imprint of increasing mobility on a globalizing world is so profound that it calls for analysis far beyond the forums of established academic disciplines". On that basis, Bergmann and Sager see need to "describe and explore contemporary and narrow notions of mobility [and] . . . investigate and nurture their transition to richer conceptual models and practices" (8).

Let me explain why such matters inform the parameters of this work by a brief return to the *0 to 100 Project*. In my reading, Nicholson *places* individuals according to age and, at the same time, he opens up *other spaces and times*—heterotopia and heterochrony. These devices allow him to consider how each person engages in journeys (voyages) and sojourns (rests), and in this way his work captures eloquently the geographies, mobilities, and rhythms of the life-course. Thus, Nicholson's photographs are among several provocations I have encountered over a number of years that prompt me to reflect more fully on the life-course. Giele and Elder (1998, 22) define studies of the life-course as those examining "a sequence of socially defined events and roles that the individual enacts over time". Such studies tend to consider the trajectories of *individual* life experiences and life spans, the influence of different periods in time, and the effects of different cohorts of age (Bynner and Wadsworth 2011; Miciukiewicz and Vigar 2013). However, the essays that comprise this work are organized as a chronology from conception to death, and do not, in fact, trace any one individual—unless my own auto-ethnographic reflections are included in that mix.

Geographies, Mobilities, Rhythms

In exploring questions of conduct and flourishing over the life-course, my strategy is geographical: concerned with space and place and their properties—including, but not limited to, dimension, direction, shape, configuration, form, boundary, structure, topography, topology, points, lines, surfaces, depths, angles, bends, folds, and arrangements. Scale is clearly implicated: in terms of measurement, proportion, calibration, size, amount, importance, rank, relative level or degree, and proportion. I am broadly indebted to synthetic work of the sort provided by Susan Marston, Richie Howitt, and Andrew Herod. Marston (2000, 238) has argued the need to redress gaps in thinking about the relationships between scale and social reproduction and consumption on the basis that "the production of spaces and places by capital supplies the context for contemporary modernity". Howitt's (2002) work on the philosopher Levinas and his contributions to geographical thought is telling for its concerns with the absent others of dispossession, and for moving beyond thinking about scale as size or level in order to consider its relationality. Herod (2010) organizes his text on scale as an analogon, starting with the body and moving out, over several chapters, to the global. In like manner, in the present work the life-course is, if you will, a scalar *heuristic*: the life-course is an idea that enables me to examine a range of stages or cohorts from conception to old age and death. It facilitates my engagement with matters microscopic and worldwide. It allows me to consider diverse geographies, mobilities, and rhythms at certain intervals over the life-course, and to inquire of each: what forms of conduct are present or indeed absent here and, singly and collectively, what can such conduct tell us about flourishing?

From *diastema* in Greek, and *intervallum* in Latin, this idea of the interval is important here, signifying extension, space, distance, an opening in time (Oxford University 1971; hereafter OED). The interval is related to the term *interstice*, which appeared in the early fifteenth century and refers to gaps, apertures, and cracks. In studies of mobility and rhythm, and of the geographies they engender, the interval—like the rest in music—marks a pause of indeterminate length in which there is space for things *to happen*

(Lefebvre 2004). And such things happen!—"vital conjunctures" as Jennifer Johnson-Hanks (2002, 878) describes them, "moments when seemingly established futures are called into question and when actors are called on to manage durations of radical uncertainty". This idea that social and spatial life is uncertain informs a significant and growing body of scholarship on mobilities from numerous disciplines (see mCenter. The Mobilities Research and Policy Center 2011). One finds, for example, work on time-and-motion studies (Sullivan 2010; Taylor 1911); design for automobility (Denfeld 1974; Wright 1945); time-space geographies (Ermarth 1992; Hägerstrand 1968); place and placelessness (Buttimer 1993; Buttimer and Seamon 1980; Relph 2000; Tuan 2001); and the political analysis of diaspora (Kasbarian 1996; Said 1993).There are also surveys of the field of mobility studies and its emergent theoretical, empirical, and methodological possibilities for social life (Bissell 2009a); paradigms and systems (Urry 2007); relational epistemologies and ontologies (Adey 2010b); and practices, spaces, and subjects (Cresswell and Merriman 2011). Jeff Malpas (2012, 226) has focused on the "concept of space as it stands in connection with time and place, making particular use of the notions of boundedness, extendedness, and emergence while also shedding light on the idea of relationality". There is, moreover, an established literature on mobile methods (DeLyser and Sui 2013; Hein, Evans, and Jones 2008; Jones, P. 2012; Merriman 2013; Murray 2009; Ohnmacht, Maksim, and Bergman 2009), and on the relationship of those methods to ethnography and auto-ethnography (Khan 2009; Sheller and Urry 2006; Vannini 2011).

Significant elements of the aforementioned works have been captured by Peter Adey et al. (2014) in a comprehensive examination of mobilities research and its genealogies, approaches, and disciplinary and philosophical effects; qualities, spaces, systems, and infrastructures; and materialities, subjects, events, methodologies, and prospects. This growth in research output and variety has also prompted several reviews of progress in thinking and practice. In one, Cresswell (2010b) refers to six aspects of the politics of mobility: motive force, velocity, rhythm, route, feeling, and friction. *Why does a person or thing move, how fast, in what rhythm, via what paths, in what emotional registers, and when and how does it stop*, he asks. Cresswell makes clear the need to be alert to the meanings, narratives, and ideological capacities of mobility; to the practices of mobility; and to the manner in which these are enfolded into diverse politics that influence our conduct and capacity to flourish. In another review, Cresswell (2010a) refers to work by several writers in allied disciplines. James Clifford (1997), for example, has questioned tendencies in anthropology to view the field and fieldwork as primarily localized and rooted. In doing so, Clifford has sought to understand diverse crisscrossing, and multiple and external connections, or different forms of mobility that may be more or less coerced. Cresswell's review focuses on Marc Augé and Manuel Castells, and their works on non-places and on spaces of flows respectively. For Augé (1995, 94), non-places include those implicating "spaces formed in relation to certain ends" including transport and transit, and the manner in which these are mediated. Castell's (2010) update on his own earlier work pinpoints key changes to the network society, among them the 2008 global financial crisis, and transformations to work, employment, communication, space, time, and social processes and organizational forms. Of Caren Kaplan's (1998) work on travel and postmodern discourses of displacement, Cresswell (2010a, 551) notes how it lays the "groundwork for a feminist embrace of mobility studies". Her labors also highlighted the need to find metaphors that liberate place from spatial fixity. Others have focused on daily mobility (Law 1999), displacement, gender, travel, and migration (Blunt 1994, 2007; Silvey 2004; Stratford 2000b), or gender, mobility, and sustainability (Hanson 2010).

Noting the existence of numerous antecedents informing John Urry's (2000) call for a new mobilities paradigm it was he who promulgated the need for it. Among his interests were bodies and effects; things as social facts, and the sensuous relationships of objects and people; flows and networks; and temporal and spatial shifts of diverse kinds related to dwelling and traveling, citizenship and diaspora, or regulation and disorder. Later, Mimi Sheller joined Urry and they issued a joint call to build this new paradigm and focus on broad-based investigations about how, why, and with what consequences we move. Sheller and Urry (2006, 208) justified their appeal for shifts in conceptual and empirical approaches to the study of mobilities on the basis that scholars had yet "to examine how the spatialities of social life presuppose (and frequently involve conflict over) both the actual and the imagined movement of people from place to place, person to person, event to event". They argued that a new mobilities paradigm should unsettle two sets of theories. The first set, which they attributed to readings of Heideggarian thought, is broadly sedentarist, and "locates bounded and authentic places or regions or nations as the fundamental basis of human identity and experience and as the basic units of social research" (208–9). The second set "departs from those that concentrate on postnational deterritorialisation processes and the end of states as containers for societies" (210).

In presenting an inventory of methods as part of their call for new theorizations of mobility, Sheller and Urry (2006, 219) have invited methodological innovation on "places of in-between-ness, cafes, amusement arcades, parks, hotels, airports, stations, motels, harbours . . . transfer points . . . new forms of 'interspace' . . . or connected presence in which various kinds of meeting-ness are held in play while on-the-move".[3] They have, in addition, delineated six bodies of theory that they view as key to advancing the field. *First* is work by Georg Simmel, who, they write, "established a broad agenda for the analysis of mobilities" (215). Simmel (2005 [1903]) did indeed make broad reference to the rhythms of modern metropolitan life, the functions of capital, and the effects on human life of the universal diffusion of pocket watches and a faster pace of life. Such observations seem to anticipate Bergmann and Sager's call, previously noted, that scholars should direct attention to the question of how to flourish in the face of what appears to be life's increasingly fast pace and complex velocities (see also Hubbard and Lilley 2004; Knox 2005; Nordbakke and Schwanen 2013). *Second*, Sheller and Urry (2006) have referred to the insights to be gained from science and technology studies that illuminate how the social is heterogeneous, and that allow focus on hybrid mobile sociotechnical systems and geographies. The implied movements in John Law's (1994) descriptions of place in an extended ethnography of Daresbury Laboratories exemplify this; so too the mobilities and, more particularly, mobilizations also evident in Bruno Latour's (2011) thinking on inscriptions. *Third* is the utility of the spatial turn in the social sciences—ideas about how we are embodied in, occupy, move through, or otherwise engage with diverse *scapes* and about how these are fluidly linked in systems of multiscalar mobility (see Appadurai 1996). Examining such ideas invites further consideration of how mobility is "a fundamental geographical facet of existence [and how] . . . the resulting ideologies of mobility become implicated in the production of mobile practices" (Cresswell 2006, 1, 21). *Fourth*, is the relevance of research on the body, affect, and emotional geographies. In this respect, consider David Bissell's (2007, 277) thoughts on animated suspension, embodiment, and the ways in which "waiting is no longer conceptualised as a dead period of stasis or stilling, or even a slower urban rhythm, but is instead alive with the potential of being other than this". Bissell (2009a, b) has also considered the connections among sociality, being-with-others, and the constitution of different spatial relations in long-distance rail travel. In turn, Gordon Waitt and colleagues have made several contributions exploring different forms of embodiment and the rhythms and mobilities that attend them (Duffy et al. 2011; Waitt and Cook 2007; Waitt and Harada 2012; Waitt, Ryan, and Farbotko 2014). *Fifth* is the relevance of work examining topologies of social networks, global connections, small worlds, mobilizing community, and government, and the effects of aeromobility (Adey 2010a; Stratford and Wells 2009). *Sixth*, Sheller and Urry (2006, 216) note that thinking about "how such complex patterns form and change will be crucial to future mobilities research as it intersects with scientific research into dynamical systems". In this regard, they refer to dynamism, disaster, and risk—for example, mobilities that implicate disease vectors (Burges-Watson and Stratford 2008) or climate change migration (Farbotko 2010b; Veitayaki 2010).

However mobilities are theorized or empirically studied, they start and end some-where and some-when, and yet unsettle existing assumptions about both space and time (Hannam, Sheller, and Urry 2006). Hence, Tim Edensor (2010b, 1) looks to Henri Lefebvre's ideas from *Rhythmanalysis*, and speculates about their worth for "investigating the patterning of a range of multiscalar temporalities—calendrical, diurnal and lunar, life-cycle, somatic, and mechanical—in which rhythms provide an important constituent of the experience and organisation of social time". Lefebvre demonstrates how rhythmanalysis affords opportunities to think about movement, rhythm, pattern, and power in terms of the temporal complexities of embodied everyday life. Introducing the English translation of *Rhythmanalysis*, Stuart Elden (2004a, vii) has noted that in "the analysis of rhythms—biological, psychological and social—Lefebvre shows the interrelation of understandings of space and time in the comprehension of everyday life. This issue of space and time is important, for here, perhaps above all, Lefebvre shows how these issues need to be thought together rather than separately". Elden gestures to Lefebvre's desire to push Henri Bergson's (1946 [1923]) ideas about time, encapsulated in specific notions about duration and described in essays in *The Creative Mind: An Introduction to Metaphysics*. There, Bergson considers duration using three metaphors. The first is of two spools—one unwinding and representing aging, the other winding up and representing the accumulation of memory. The second is the spectrum—suggestive of a multiplicity of changing shades of experience over a life span. The third metaphor is a tiny piece of contracted elastic that may then be extended with a view of focusing not on the line, but on the action that is traced—again, evocative of the idea of a life-course. The cumulative effect of these three metaphors is that duration is seen as heterogeneous, mobile, interwoven, ceaselessly changing, and embodied. Indeed, as Suzanne Guerlac (2006, 4) notes in this regard:

> Bergson enables us to return to questions associated with temporality, affect, agency, and embodiment that were bracketed within the structuralist/post-structuralist context . . . Reading Bergson enables a reengagement with the concreteness of the real . . . without sacrificing the perspective that drove the critique of representation and of the unified subject of consciousness.

However, Elden (2004a) suggests that Lefebvre seeks to focus not on duration, but on the instant or moment (Nietzsche's idea of the *Augenblick* or blink of an eye). Reading Lefebvre (2004) himself affirms that view: he sees the body as the point of contact between biological and social rhythms—a first point of analysis, a tool, a metronome. He conjectures the existence of several kinds of rhythm: polyrhythmia (many rhythms), eurhythmia (harmonious rhythms), arrhythmia (disjointed rhythms), and isorhythmia (equal rhythms). These categories of possibility force one to register the *stillness of nothing* in the world. Lefebvre's categories of rhythm also enable consideration of the ways in which "a new focus on mobilities in geography allows us to re-centre it in the discipline" in order to unsettle prevailing ideas about geographical knowledge comprising "a world of places and boundaries and territories rooted in time and bounded in space" (Cresswell and Merriman 2011, 4). Importantly, all such changes will be experienced and mapped in and through the body as rhythms. As Lefebvre (2004, 15) has suggested in *Rhythmanalysis*:

Everywhere where there is interaction between a place, a time and an expenditure of energy, there is rhythm. Therefore: a) repetition (of movements, gestures, action, situations, differences); b) interferences of linear processes and cyclical processes; c) birth, growth, peak, then decline and end.

Lefebvre's (2004) analysis of rhythms first concerns what he calls "secret" rhythms. Such movements and temporalities are not unknowable; rather they do not "publicize" themselves: "Do not confuse silence with secrets!" he writes (17). Secret rhythms, then, are not inaccessible or beyond knowing; but one must work to apprehend them. Think of the *0 to 100 Project*, of young Robin Walker, and of the partly secreted *motilities* of his conception and gestation. Other rhythms that concern Lefebvre relate to psychology, including recollection and memory, such as that exhibited by Alma, another of the *0 to 100* participants: "How old am I? Wait'll I think now . . . it's been a long time since I thought about how old I was . . . Um [pause] my . . . phew . . . I am [pause] give me a piece of paper . . . yeah, I'm 90, at least. The best thing about being 90 is that you don't have to worry about it!". Lefebvre's analysis then traces a range of public and social rhythms, such as those that constitute celebrations or fêtes, or that give expression to fatigue. Third are fictional rhythms, among which Lefebvre names elegance, gestures, learning processes, and the imaginary—one assumes on the basis of an understanding that *fiction* is etymologically allied to the work of fashioning or of transformation. Last are rhythms of domination that may exist in speech, for example, and that aim for effects beyond themselves.

Constituting his ideas as a method, Lefebvre suggests that the "rhythmanalyst will . . . be attentive . . . will listen to the world, and above all to what are disdainfully called noises, which are said without meaning, and to murmurs [*remeurs*], full of meaning—and finally . . . will listen to silences" (19). Neither passivity nor asymmetry will be hallmarks of the rhythmanalyst's methodological obligations. Rather, the body is to serve as a measure, a metronome, tracking everything that moves (and *everything* moves) without "privileging any to the detriment of any other . . . [immersed in] this tissue of the lived, of the everyday [and] . . . arriving at the concrete through experience" (20–21). Furthermore, the imaginary or simulacra that is the *present* is rejected in favor of an authentic *presence*: a "dramatic becoming . . . [an] ensemble" of multiple presents (23). Lefebvre argues that this authentic presencing challenges our imprisonment in the ideology of "the thing". This ideological incarceration is something Lefebvre particularly wants to disturb in order to reinstate the "sensible into consciousness and in thought . . . [and] accomplish a tiny part of the revolutionary transformation of this world and this society in decline" (26). Thus, Lefebvre's intentions to outline a method of approach to understanding everyday life are unambiguously part of an associated desire to work for transformations: each of the rhythms that he outlines may have "ethical, which is to say practical, implications" (18). In this respect, Elden (2004a, x) usefully reminds the reader that *Rhythmanalysis* is the fourth and final offering in Lefebvre's occasional series, *Critique of Everyday Life*, and represents "a corrective to Marxism's tendency to privilege time over space", which is also exemplified in *The Production of Space* (Lefebvre 1991).

On Conduct and Flourishing

In the essays that follow, I take different and chronologically ordered stages of the life-course and consider the geographies, mobilities, and rhythms that characterize several significant problems that interest me. Some of those problems, I have worked on over several years; some are new. I elaborate on each in more detail below, but summarize them now so that the ensuing discussion on conduct and flourishing makes sense. These problems or cases are, if you will, the warp threads of a larger tapestry mapping the adventures and misadventures—the vital conjunctures—that typify our lives. Running across each chapter are weft threads in the form of two questions. How have we and how do we conduct ourselves? And, in order to flourish and create the conditions for flourishing to spread, how might we conduct ourselves differently? The six cases that help me to get at these questions are as follows: (a) the figuration of the pre-embryo, embryo, and fetus, and questions about the geographies, mobilities, and rhythms of bringing to life; (b) the possibilities of homelessness, statelessness, and forced mobility for young island children, and the consideration of territory, sovereignty, and citizenship under climate change; (c) life for teenagers and younger adults who seek—not without challenge—to use the spaces that constitute settlements in ways they deem playful and others deem disrupting; (d) life for adults engaged in commuting, that 'daily grind' wherein we try to dwell-in-motion, and which is increasingly subject to acts of violence; (e) life for older adults coming to terms with the changing nature of embodiment by developing and maintaining fitness and well-being; and (f) life for the oldest old, and the final journeys that they make to what Shakespeare termed "the undiscovered country".

For me, questions of conduct and flourishing implicate Michel Foucault and others indebted to his work. There has been some nervousness on my part in juxtaposing Lefebvre and Foucault. Eleonore Kofman and Elizabeth Lebas (1996) document the ways in which Lefebvre was committed to demonstrating the conceptual links between Marx and Nietzsche and the idea that revolutionary change necessitates transformations of the self. In the process, they note that Lefebvre was critical of what he saw as Foucault's partial explanations and tendency to privilege systematized knowledge (*savoir*) at the expense of the experiential (*connaissance*), which is evident in the ways in which the former understands dressage, the latter discipline. Kofman and Lebas also comment on the manner in which Lefebvre conceived of difference as emerging from lived struggles rather than "particularity, originality or individualism" (26). However, some of my initial apprehension has been moderated by the point that Lefebvre thought Foucault's use of "historical material to support a philosophical project, an attack on the Western *logos*" was powerful (Elden 2004b, 239). Edward Soja (2009, 18) certainly emphasizes that the philosophers' "co-presence in Paris . . . is undeniable and so too is their nearly simultaneous development of essentially similar ideas about the ontological significance of space and the powerful forces that emanate from the spatiality of human life" (see also Crampton and Elden 2007; Goonewardena 2011). On the one hand, Lefebvre (2004) provides clear guidance on methods of approach to rhythmanalysis, particularly in comments on dressage and the chapter "Seen from the Window". In that essay, he recommends that the rhythmanalyst deploys multiple senses of observation in order to understand "the rhythmic lineaments of everyday life [that] are weighted with power" (Edensor 2010b, 8). On the other hand, in numerous papers on the anatomo-politics of human bodies, the biopolitics of populations, governmental regimes, and heterotopia, several significant insights on conduct are provided by Foucault (1976, 1980, 1982, 1986, 1991). Taking up that work, Nikolas Rose (1998, 178) asks of governing, conduct, and power: "Who speaks, according to what criteria of truth, from what places, in what relations, acting in what ways, supported by what habits, routines, authorized in what ways, in what spaces and places, and under what forms of persuasion"? Mitchell Dean (1999, 17–18), too, observes that "we govern . . . others and ourselves according to various truths about our existence and nature as human beings . . . the ways in which we govern and conduct ourselves gives rise to different ways of producing truth".

These several strands inform my thinking about how life's potential and our capacity to flourish are wrapped up in varied rhythms, mobilities, and geographies, and I shall have more to say about flourishing in the pages that follow. Here, and only briefly, I tease out a few threads of this weft line of thinking. My point of departure has been my reading of *The Nicomachean Ethics*, and of Aristotle's (350 BCE(b)) understanding of *eudaimonia* (well-being, virtue, flourishing). He asserts that everything we think, all of our actions, and all of our arts (practices) are intended for some higher purpose or some good; he contends that this end is, in itself, "the chief good" (Book I:1). For Aristotle, *eudaimonia* moves well beyond adventitious or gratuitous pleasure, and is implicit in an honorable, functionally useful, and complete life.[4] That life, throughout and in total, is underpinned by prosperity or good fortune, which Aristotle understands as partly comprising external goods without which "it is impossible, or not easy, to do noble acts" (Book I:8). These goods include friends, riches, political power, good birth, worthy children, and beauty; but neither unjust nor grasping conduct is countenanced (Book V:1), and nor is a propensity to excess (Book X:8). Throughout the *Ethics*, Aristotle avers that by "human virtue we mean not that of the body but that of the soul; and happiness also we can call an activity of the soul" (Book I:13). Nevertheless, *eudaimonia* is further enriched by providing for oneself, "for our nature is not self-sufficient for the purpose of contemplation [a mode of conduct most highly valued], but our body also must be healthy and must have food and other attention" (350 BCE(a), Book X:8). Aristotle elaborates on the relationships among the 'nutritive soul', the body, movement, and the nature of being in *De Anima*, in which he writes that "it is obvious that the affections of soul are enmattered formulable essences" (350 BCE(a), Book I:1). In other words, the soul is not separate from the physical body and is its first actuality, purpose, or final cause.

Returning to *The Ethics*, a flourishing life requires the exercise of practical wisdom (*phronesis*), which is the form of judgment that aids and enhances particular moral virtues. Training and habituation are key to such a life, and through them virtue may be won by "a certain kind of study and care [because to] entrust to chance what is greatest and most noble would be a very defective arrangement" (Book I:9). Importantly for my purposes, in *The Ethics* Aristotle argues that the prospect that "the fortunes of descendants and of all a man's friends should not affect his happiness at all seems a very unfriendly doctrine" (Book I:11). In other words, conduct and flourishing are ontological at multiple temporal (and spatial) scales, for in Aristotle's thinking, the ancestors want to know that their ways of being in the world had good effects into the future.

Let me try and weave these threads together in terms of thinking about the possibility of imagining other times and other spaces, and linking that possibility to modern understandings of prosperity. For this work, I turn briefly to Erik Olin Wright (2010, 2013), who has questioned a widespread resignation to the seeming and total inevitability of advanced liberal capitalism; and has sought to systematize real, social, and emancipatory alternatives to foster vibrant alternatives. His larger premise—that deficits in human flourishing are socially caused, and therefore resolvable—does not prevent his appreciation that 'the road to hell is paved with good intentions'. At the same time, Wright acknowledges the paradox of conjoining the terms 'real' and 'utopia', which after all is a pun for both good place (*eu-topia*) and no place (*ou-topia*). Nevertheless, he asserts the possibility, now, of flourishing by fostering equality, democracy, and sustainability. His four-step agenda for change involves identifying the moral principles that may be used to judge the worth of social institutions, then applying those principles to diagnose and critique our present institutions in order to inform the creation of viable alternatives and advance frameworks by which to transform social life and realize those alternatives. These tasks, he writes, are like a voyage in which one first maps what is wrong in the world; second, determines what then motivates a need to depart from it; third, imagines what new world is sought; and fourth, works out, practically, how to move from one to the other.

For Wright (2010, 2013), flourishing is central to the journey, both as a foundational principle and as an end. For this present work, which has geography as its home terrain, the metaphor of the journey is apt. As Paul Cloke (2002, 588) has observed, "questions about living ethically and acting politically as human geographers are integrally wrapped up in the life

experiences of the individual", and require critical perspectives and engagement with moral questions. While often despondent about the state of the
world, Cloke asserts an optimistic outlook overall; this on multiple bases
that there are increasing numbers of examples of both "geographically sensitive ethics, and an ethically sensitive geography" (591). There are also
instances of productive and normative self-criticality; deep-seated concern
among scholars to improve the lives of other; and a growing appreciation
of the interconnectedness of cultural, material, political, and economic
systems and processes. There is a related understanding that geographers
are among those in the humanities and social sciences who have important
roles in reading and resolving the problematic tensions between identity and
socioeconomic redistribution—which might be read as securing prosperity.
Finally, there is an apparent willingness to "take seriously the notion of
evil" (592).

Does talk of flourishing necessitate consideration of evil, and what is
meant here by that term? Cloke's (2002) own musings on this subject are
drawn, at least in part, from an essay by Yi-Fu Tuan sketching out the
relationship between geography and evil. As Tuan (1999, 106) observes,
"human geography is a well-established field, and people certainly have
feelings—they love, hate, build, destroy and kill—and yet, until well into the
second half of the twentieth century, geographers have managed to avoid
morals and morality altogether, or skirt around their edges". Noting that
the idea of evil is largely absent from the discipline's lexicon, and is a term
also avoided by many moral philosophers, Tuan nevertheless asserts evil's
undeniable facticity. On this basis, he examines our propensity for destruc-
tiveness, cruelty, sadomasochism (that is, relations of domination and
submission), and compartmentalization (forgetting, disconnecting, and dissociating). Tuan concludes that these themes are familiar stuff for humanists
and should also be for geographers.

It is noteworthy, I think, that only in the eighteenth century did the word
evil come to be broadly associated with 'extreme moral wickedness'. Prior
to that time, it tended to be used as an adjective to mean bad, cruel, unskill-
ful, or defective; and as nominating harm, crime, misfortune, or disease
as forms of conduct and conditions of being that designate an absence of
flourishing (OED). It is these shifting moral terrains that have occupied my
attention in the essays to follow, and it is those I now seek to describe in
more detail.

MAP AND COMPASS

Shifting Places of Origin

Chapter two rests on an apparently unremarkable observation that the
body's interiors have their own processes, rhythms, movements, and—in
and through the boundary object of the skin—complex geographies; I say

apparently here in recognition that a problematic binary is invoked by this phrasing. Among such bodily processes are those potentially generative of life: ovulation, ejaculation, and conception. These processes typically involve others' interiors, but may implicate pipettes or test-tubes and other environs.

How might one think through the rhythms, mobilities, and geographies involved in bringing to life? This question is important for at least three reasons. *First*, it requires consideration of the idea of the interspace referred to earlier in relation to Sheller and Urry's (2006) methodological agenda. Here, however, the focus is not on the aforementioned lounges, cafes, airports, stations, motels, or harbors, but on the interior of the body, which has occupied only some scholarly attention in mobilities research. In the second chapter, there is a lengthy discussion about human embryology and morphogenesis, followed by an elaboration on the ways in which these have come to be pictured. *Second*, questions about the rhythms and the ensuing mobilities and geographies involved in bringing to life raise other queries about the embryo and fetus, and about how they, pregnant bodies, or petri dishes and other technologies are permitted to move into or out of particular spaces that then are constituted in moral or ethical terms. For instance, reporting on progress in scholarship on calculation and territory, and by reference to research by Robyn Longhurst (2006), Jeremy Crampton (2011, 97) has asked "how do bodies interplay with boundaries"? Longhurst's work, informed by ideas of abjection (that is, mobilizing acts to expel or to move away from that which is repugnant) and moral geographies is centered on a woman in New Zealand whom Longhurst refers to as Nikki. It was Nikki's quest to work with pornographic film maker Steve Crow to document her sex life while pregnant, and to record the birth of her child as a final climactic event. According to Longhurst (2006, 218), at "stake in the making and screening of pornographic films about pregnancy and birth are questions about the conceptualization of pregnant, birthing, and babies' bodies. When birthing the borders of a woman's body change rapidly—they are not fixed—but rather have volatility". My own focus is on other volatilities involving ova, sperm, and the figuration of life before birth. *Third*, then, questions about the geographies, mobilities, and rhythms of bringing into life prompt still other inquiries about conduct and flourishing; for example, those about fetal personhood, parthenogenesis, or somatic cell nuclear transfer experiments including human therapeutic cloning. Such matters touch deeply on what I have come to think of as places of origin and exile.

Fluid Terrain

Some six hundred and fifty million people live on islands and archipelagos. Islands are often constituted as the frontline of climate change and sea level rise. Indeed, the prospect that islands may be part of ground zero for such change prompts varied responses. Consider voyeuristic forms of disaster tourism (Farbotko 2010a), or local reactions to the possibility of forced migration (Paton and Fairbairn-Dunlop 2010), or the reinscription of continental Canada as an archipelago if or as the Arctic thaws (Vannini et al. 2009).

The forecast of rising tides invites consideration of Philip Steinberg's (2005) work on the Portolan charts of the Renaissance Mediterranean. In the charts, Steinberg argues, islands signify a flawless relationship between physical and political geographies—with all the implications which that idealization has for post-Westphalian notions of sovereignty and territory. Understood as elemental spaces, in Portolan charts islands were conceptualized as simultaneously solid and mobile—of the sea rather than of the land. So, too, Elizabeth DeLoughrey (2007, 3) suggests of island space that paying attention to movement "offers a paradigm of rooted routes, of a mobile, flexible and voyaging subject who is not physically or culturally circumscribed by the terrestrial boundaries of island space". In present times, subjects apparently freed from such circumscriptions may choose, or in fact may be forced, to leave those bounded spaces to find work in continental or mainland spaces and return remittances to those left behind (Bertram 2006). They may be incorporated into the infrastructures of island tourist enclaves that are "disembedded from national territories" (Sheller 2009, 1386). Or they may experience the need to migrate on the basis of loss of territory, nationhood, and sovereignty as a result of sea level rise (Yamamoto and Esteban 2010).

It is particularly concerning to me that in much work on islands, my own included,[5] the island child is sometimes rendered silent, absent, or Other (but see Blake 2009; Dryden, Arata, and Massie 2010; Wolf 2006). In chapter three, I want to unsettle this limited consideration, while at the same time pointing out an ostensibly continental focus in mobilities research. How might one think about the geographies, mobilities, and rhythms of anthropogenic climate change in relation to young islanders, questions of citizenship, and ideas about flourishing? I seek to give voice to that question by reference to *Fresh! A Map of a Dream of the Future* (Low et al. 2010), a project involving academics, educators, artists, and young Tasmanians that sought to elicit from the last of these their views on climate change, resilience, and island life. The project and third chapter position children as political actors, and respond to Tracey Skelton's (2010, 146–7) observation that political geography is "well placed to open its borders and analyse the political roles young people play in society. This should not just be about the effect of power on young people, but also the political power young people wield through their practices, resistance, strategies and challenges".

Grind and Trace

Gill Valentine, Tracey Skelton, and Deborah Chambers (1998, 7) argue that the part which young people "play in all our geographies" necessitates understanding their experiences. In chapter four, the focus is upon older adolescents and young adults, on skateboarding and parkour, and on the possibility that their conduct invites consideration of the worth of playful spaces and of generosity. I began research and policy work on skateboarding in 1999, urged on by such possibilities and by observing the vilification of young skaters in the city in which I live and elsewhere. Over the intervening period, considerable research has been done on skating and parkour as these concern young people's marginalization (Guss 2011; Jenson, Swords, and Jeffries 2012); the use of space to resist and normalize (Rawlinson and Guaralda 2011; Stratford 2002); strategies of legal and regulatory containment (Carr 2010; Chow 2010); and the paradox that modern culture is overwhelmingly about youth consumerism whereas the governance of public spaces problematizes the presence of adolescents in them (Daskalaki, Stara, and Imas 2008; Howell 2008). This inconsistency arises, perhaps, because puberty "is an ambiguous zone within which the child/adult boundary can be variously located according to who is doing the categorising" (Sibley 1995, 34). Either way, many such categorizations produce deficit models in which young people are constituted as fringe-dwellers, delinquent, unruly, unproductive, and resistant (Scott and Steinberg 2008; Stevens et al. 2007).

Walking, we know, can be a form of resistance (Anderson 2004; Edensor 2010c; Macauley 2000; Mauss 1973; Morris 2004; Pink 2007), and considerations of it so conceived and practiced provide new insights on skating and parkour. Among the most celebrated of offerings is Michel de Certeau's essay "Walking the City". Having serenaded the virtues of gazing from the World Trade Center, de Certeau (1984, 93) observed that:

> The ordinary practitioners of the city live "down below", below the thresholds at which visibility begins. They walk—an elementary form of this experience of the city; they are walkers, *Wandersmänner*, whose bodies follow the thicks and thins of an urban "text" they write without being able to read it. These practitioners make use of spaces that cannot be seen; their knowledge of them is as blind as that of lovers in each other's arms . . .

Not so with skaters, *traceurs* and *traceuses*[6] for whom feet, eyes, hands, proprioceptors, and other parts of thinking, sensing body-selves are used to access spaces that often cannot be seen in two senses of the phrase. First, they use fixtures, sites, or spaces such as banisters or wall edges that others *can* see but do not *read* as navigable; second, they access spaces such as rooftops that others might imagine to be present in space, but cannot see because they do not (or hardly ever) encounter them in their own perambulations.

These practices of skating and tracing invite close readings of bodies rhythmically moving on streets, footpaths or sidewalks, in parks, on infrastructure. They also prompt consideration of intensity and density of affect for those who engage in skating and parkour and for those who do not.

Chapter four, then, is concerned with several questions: how might existing scholarship on walking, for example, inform a reading of skating and parkour and their geographies, mobilities, and rhythms? How do older adolescent and young adult skaters and traceurs use urban spaces and produce ludic geographies contoured by particular movements and rhythms? How and with what effect on flourishing are their practices entangled in specific modes of conduct and understandings of settlements that appear to render them unwelcome?

Encountering the Circle Line

Adulthood is a long stretch, and three chapters in this volume examine different rhythms, mobilities, and geographies of these decades. Chapter five, the first of these, explores commuting as a mundane venture in the intervals of adult lives—a traveling between, a form of conduct—that simultaneously is regulated, regularized, and at risk of disruption—sometimes by dint of force and in ways that violently affect our capacity to flourish. Now, I recognize that children are implicated in the instrumental functions of commuting to school and to work. Commuting certainly *is* intergenerational. (That fact was brought home to me in October 2011 on a Piccadilly Line train to Heathrow. There, I had an animated conversation with a six-year-old about the music of Bob Marley, about whom she had learned in class that day. Thrown together for a while in the over-packed train, with her waxing lyrical about reggae music, we were quietly watched by her grandfather, who accompanies her brother and her to school each day.)

In addition, I acknowledge that scholarship on children's commuting patterns ranges from traditional transport studies to those involving more experimental tones. For instance, one would be hard-pressed to find greater difference than that between Yeung, Wearing, and Hills' (2008) positivist study of school students' transport practices as active or passive, and Kullman and Palludan's (2011) rhythmanalytic explorations of the school journeys of five to seven-year-old residents of Copenhagen and Helsinki. Such different forms of scholarship have evoked considered responses from at least two quarters. First, between transport geography and mobilities research, Shaw and Hesse (2010) illuminate certain points of tension and others of conceivable rapprochement. They note that some of the nascent agendas for mobilities research are to be found in Edward Ullman's work from the 1950s. Among human geographers' interests, for example, Ullman (1953, 54, 56, 58) identified "transport routes and flow . . . reciprocal relations and flows of all kinds among industries, raw materials, markets, culture, and transportation . . . [and] mapping and analyzing the flows of goods and peoples in the area—a kinetic or dynamic aspect of geography". Referring to another of Ullman's essays from 1954 entitled "Geography as spatial interaction", Shaw and Hesse (2010, 306) maintain the relevance of his work by considering the ways that place "is bound up in a network of interaction and interdependency". They further propose that geographers engaged in transport geography and mobilities research "would do well to further advance mobility as a foundation concept in geography . . . [given that] in its most basic form it is about movement between places across space" (308), often in terms of a daily journey. Second, Cresswell (2010a, 554) has identified the need for a bridge between transport geography and mobilities research, suggesting that we attend to "the ways in which travel time is filled with significance". In thinking about how adults experience commuting, and in seeking to take up Cresswell's suggestion, in chapter five I settle upon the London Underground, a far cry from my own solitary ten-minute drive from home to the University of Tasmania via North Hobart's Raincheck Café. Not of, but enamored with London, in recent years I have become a regular visitor, gradually becoming familiar with the Tube to a point where my demeanor might have a patina of confidence; sometimes I am asked for assistance by tourists. Does that make me a commuter when there? Certainly, as part of an exploration of Lefebvre's (2004) notions of presence, I have gotten into the habit of "doing" the Circle Line each time I am in London. My journeys are about being there, planted and mobile, practicing an embodied presence, and watching and considering others' mobilities and rhythms (see also Bassoli et al. 2007; Cronin 2008). That work also enables a more general consideration of the adult body in the labor force, the daily "grind" of commuting, and different modes of being and doing. Those encounters have prompted three questions that I pursue in chapter five: how do adults experience the practice of commuting? In what ways might commuting be considered dwelling-in-motion? What do increasingly violent disruptions to commuting as dwelling-in-motion mean for how we think about conduct and flourishing?

Move It or Lose It

Attention shifts in chapter six to a reflection on the geographies, mobilities, and rhythms of the middle-aged and older adult body. It is said that we are what we eat; equally, we are how we move. Foucault (1975) has written about exercise as a regime of practice in the disciplining biopolitics of the population. Others, such as Lois McNay (1992), added to his many insights the important corrective of gendered readings. The idea that exercise may transform body and society is not new. Take the twenty-fifth anniversary publication of the Australian Health Society, which was active in the late nineteenth and early twentieth centuries. There it was observed that the closing quarter of the former century was to be distinguished for the variety and magnitude of its contributions to the great "departments" of public and personal hygiene, and the advances of modern science.

> Thus to the individual have been brought home the supreme importance of heredity, the true position of diet, the value of exercise, the means

and limitations of education, the interdependence of the physical, the mental, and the moral, the disease relationships of man with plants and animals, and the new world of friends and foes that exist in Germland, with information about the prevention of hostile invasion, victory over or immunity against attack, rout at points of exit, and destruction in breeding grounds. (Australian Health Society 1900, 4)

This celebratory synopsis is replete with ideas about traveling (being brought home); the moving (into position) of sensible eating habits and of bodies trained to work in particular ways; the vectors of disease; the displacement (shifting out) of ideas about miasma with those about germ theory; and so on (see, for example, Stratford 1998a; Stratford 1998c). Moreover, the passage implicates particular forms of conduct—diverse means by which to assemble the self—and ideas about what it means to flourish.

Granted, contexts change, yet similar concerns exist in the present and, acknowledging their multi-generational impacts, I am interested in understanding how the idea of staying young—or of being not-old—is taken up among those in middle and later life in ways that explicitly involve moving. A search of Internet sites such as Real Age (Live Life to the Youngest) reveals the top ten of an estimated fourteen hundred pages on youthfulness; among these are headlines on best and worst cities for staying young, how to keep staying young, and move it or lose it. Underpinning such efforts by changing one's daily rhythms and patterns of mobility are several narratives of the self—not least among them growing trepidation about aging as a decline. In this regard, in one study using auto-photography as a method, participants engage in three narratives about their identities that are oriented to health, performance, and relationships. The study's author, Cassandra Phoenix (2010, 167), argues that in "combination, these . . . offer insight into . . . the meaning of aging, and also act as counter-narratives to dominant narratives of decline in older age". Other research by Julia Rozanova (2010) examines the idea of staying young as it gains currency in Canada's national *Globe and Mail* newspaper. She, too, points towards forms of identity, suggesting that these are based around neoliberal principles of individual choice (to be healthy), individual responsibility (for not being so), and individual forms of conduct to age successfully by staying engaged.

In light of these observations, in chapter six I want to ask how is the aging body constituted, especially in terms of decline and increasing incompetence? How have such notions been challenged, particularly by counter-narratives of positive aging? How are certain geographies, mobilities, and rhythms implicated in regimes to produce fitness as an alternative to decline? What do these narratives reveal about what it means to flourish in middle and later life? How and to what effects might "moving it" constitute interest in dressage and training, two forms of conduct which concerned Lefebvre and Foucault, among others?

The Undiscovered Country[7]

In *Deathscapes. Spaces for Death, Dying, Mourning and Remembrance,* Avril Maddrell and James Sidaway (2010) observe that the works in their collection add new perspectives on space and place via reflections on loss, remembrance, identity, memorialization, emotion, and affect. The geographies, mobilities, and rhythms that attend death, the undiscovered country, are the subjects of chapter seven. Again, embodiment is key. Kirsten Simonsen (2003, 165) refers to the idea that the body is a "fundamental situation" of lived experience, and underscores that this experiential situatedness is both energized and spatialized (after Lefebvre 1991, 195). Senescence and dying, however, stand out as stages where such energies and their rhythms dissipate. Those processes of declining energy may be sudden and rapid, or come on subtly and take decades, even among the oldest-old. Some will be robust until shortly before death; some might have been in a vegetative state for years but "tick on"—literally—because there is a pace-maker in their chest cavity.

Acknowledging Simonsen's (2003) work, there is limited scholarship on *rhythmanalysis* and aging, in contrast to a proliferation of work on how mobility amongst elders is managed, diminishes, or disappears (Andrews et al. 2006; Gilleard and Higgs 2011; Kenner 2008; Lord, Després, and Ramadier 2011; Phoenix and Grant 2009; Temelová and Novák 2011). Perhaps that is because, as Heinz Kaiser (2009, 411) indicates, mobility is "a constitutive and essential element" in the quality of life of elderly people. Mobility is used as a signifier for the extent to which elders are able to flourish—for example, affecting whether they feel constrained by public and private transport systems that account poorly for their needs, or sense that they are trapped inside bodies that tend to arrhythmia, restriction, and shrinking scale and range. In this regard, as Laura Hurd Clarke and Alexandra Korochenko (2011, 495) observe, the body is the site upon and in which "we most immediately experience the social and physical realities of growing older . . . all of our experiences of aging are invariably embodied".

I have the great fortune to belong to a community in which I have regular interactions with several fully cognizant and mobile men and women in their late eighties and nineties. As I wrote the first draft of this introduction, I received word that one had lapsed into a coma from which she was not expected to emerge. I had known her since she was eighty-eight and, in the intervening five years, listened carefully to her incisive comments on wide-ranging subjects from education to peace and politics to gardening. I watched with increasing vigilance as she moved from independent walking to a stick to a wheeled frame. As this cumulative 'degeneration' occurred, at our regular gatherings I began to place a footstool at the base of the chair she favored. I listened more actively as her thoughts—which had always been profound—began to wander just a little. In retrospect, I had been the quiet observer of the mobilization of the end of days; it is not the first time in my experience, and I daresay it will not be the last. Her journey to the undiscovered country, as Shakespeare would have it, was not sudden, dramatic, or violent. When she was gone, the course that her life had taken was celebrated by those who remained; she remains an absent-presence in our gatherings. My task in the penultimate chapter, then, is to consider senescence, dying, and death among the oldest-old (Froggatt, Hockley, and Parker 2010; Leavy 2011; C. Nicholson and Hockley 2011). How are we to understand the movement through stages of liveliness to frailty and dying into death among the oldest-old? How are we to reflect on the rhythms, mobilities, and geographies of life at its close?

Space to Flourish

In the last chapter, I return to my initial points of departure: questions about the geographies, mobilities, and rhythms of the life-course which allow reflection upon questions of conduct and the manner in which we constitute the spaces of a good, just, and flourishing life. I provide a comprehensive summary of each chapter of this volume, drawing out what I see as its particular insights, and forging new links across the warp and weft of the things that have concerned me. And finally, I make reference to alternative readings of Aristotle's ideas about *eudaimonia* and what it might mean to live a flourishing and complete life. In doing so, I posit the importance of remembering that we have immense capacity to be *homo reparans*—a repairing and caring species—and that this capacity is enlivened *because* we are placed in the world as vibrantly spatial, mobile, and rhythmic creatures.

NOTES

1. Throughout the chapters that follow, I come back to the participants in Nicholson's (2011) project, from time to time quoting what they shared with him about their ages and lives. All such quotes may be located on the *0 to 100 Project* website and app.
2. I have written this passage assuming that none of the participants in the *0 to 100 Project* is an Indigenous person, but I have no evidence to corroborate that one way or the other. The point, I think, stands.
3. Note, however, Shaw and Hesse's (2010, 309) response to that call: "In our view this [claim to methodological innovation] merits further critical reflection, for surely what is at stake is only the tweaking of particular methods capable of harnessing the power of existing methodologies in mobile situations". I would venture to suggest that more than tweaking is both possible and warranted.
4. Aristotle has been charged with dismissing the capacity for the young, slaves, and women, to engage in the sort of intellectually and morally virtuous political life that he seeks to constitute as eudaimonic. Yet, as Hollie Mann (2012, 201) has suggested, reading Aristotle's elitism becomes less clear-cut, and spaces for emancipation become conceivable if one accounts for and finds the potential correctives in his:

association of the love mothers feel for their children not with some essential female quality but rather with the work of laboring to produce another human being. His point is not the essentialist one that women are more loving toward their children simply because they are women, though this passage is sometimes interpreted this way. Rather, Aristotle is saying that mothers come to love children more than their fathers only as a result of the activity they have done for them, in particular, pregnancy, labor, and, perhaps we can infer, the work of child rearing more generally. Such a reading perhaps requires that we take seriously the claims of Jill Frank. She has persuasively argued that we look beyond Aristotle's direct claims about the inferiority of slaves and women and consider more closely his fluid, constructivist conception of nature, which Aristotle clearly believes is contingent on one's function, and one's function is contingent on the activity that one takes up, or is allowed to take up.

5. Living on an island for nearly two decades, and being a geographer, I have been drawn to study matters of space, place, and mobility in island life. That work has been expressed in a number of papers on international governance for oceans, coasts, and islands; intergovernmental relations, and governmental approaches to community and development; and sense of identity and belonging (Stratford 2003, 2004, 2006a, 2006b, 2008, 2009; Stratford, Armstrong, and Jaskolski 2003; Stratford et al. 2011). Some of that work has, in fact, engaged children, as I report in chapter three.

6. *Traceur*, literally bullet, is French masculine for a practitioner of parkour, itself a neologism based on the term *parcours*, meaning obstacle course; *traceuse* is the feminine form. Like skaters, most practitioners are males.

7. The relevant passage is:

> . . . Who would Fardels [burdens] bear,
> To grunt and sweat under a weary life,
> But that the dread of something after death,
> The undiscovered Country, from whose bourn
> No Traveler returns . . . (Shakespeare c. 1600, *The Tragedy of Hamlet, Prince of Denmark* Act III, Scene I)

2 Shifting Places of Origin

In considering the geographies of relatedness, Catherine Nash (2005, 450) gives critical consideration to the "social organization of the 'natural facts' of sex and reproduction" and to the naturalization and geopolitical organization of nation-states via the family. Her analysis embraces transnational adoption schemes and gamete transportation for assisted reproduction, and the curtailment of immigration on the basis of arguments about 'balancing' ethnic mix. At the same time, Nash argues for more work by geographers on the effects of biotechnologies that "create new understandings of embodiment, subjectivity, sociality, the human and human genetic diversity and relatedness" (450). On the understanding that human reproduction is both primal and foundational, she seeks to demonstrate how geographies of relatedness foreground the links between the intimate and institutional, and subjectivity and governance.

In that project, Nash emphasizes the importance of our sense of "place of origin". Here, I want to examine and extend this idea on the understanding that origins and emplacement are notions and realities *that move* (which is not to suggest that Nash thinks otherwise). In this regard, I am indebted to ideas about how relational mobilities are moored or positioned, and to the notion that identity is a "process of continuous departure" (Adey 2010b, 22–3, 25, after Probyn). My aim is to ask *how might one think through the rhythms, mobilities, and geographies involved in bringing to life?* I mean to interrogate both normalizing narratives that describe conception and gestation, and other dynamics that are possible from the start of life *that may not lead to birth*. Specifically, after outlining some of the boundaries that delineate understandings of the pre-embryo, embryo, and fetus, I contextualize and then probe the meanings and significance of *The Miracle of Life*. This sixty-minute documentary was released in 1983 for NOVA, a science series produced under the auspices of a Boston-based non-commercial educational television station, WGBH–TV, a member of the Public Broadcasting Service of the United States of America (NOVA 2013).[1] The documentary gives expression to hormonal, sexual, and social processes leading to conception, and then more superficially traces stages of gestational development and birth (Nilsson, Erikson, and Lofman 1982/1986). Juxtaposed against the narratives and images of this influential origin story are other rhythms, mobilities, and geographies that arise from bringing to life in other ways. I seek to account for these and to generate insights into questions of conduct and flourishing, whose overarching importance to this work was introduced in chapter one.

BOUNDING THE EMBRYO

George Linius Streeter led the study of human embryology from 1917 to 1940, in particular as Director of the Carnegie Institution of Washington. According to his biographer, George Corner (1954, 265), it "was the urge to see and understand, not merely theorize about the unknown transition stages of morphogenesis—the beginnings of form—that made Streeter a great descriptive embryologist" and resulted in the publication of his 'horizons in human development'. These data and the pictorial evidence that accompanies them are so precise they enable the determination of "gestation age from physical characteristics in the earlier stages of development, when ordinary dimensions of weight and linear size are not helpful" (266).

Streeter's stages mark numerous developmental distinctions. Fertilization itself depends on diverse hormonal, physical, social, and scientific rhythms. There is, for example, the regular recurrence of ovulation inside the body that, in Lefebvre's (2004) terms, might be read as eurhythmic—deemed harmonious because it is 'natural'. Then there are whirring, isorhythmic pulses that emanate from machines used to aspirate and flush ovaries, and move ova into vials. And then there are other biochemical and biophysical actions that propel sperm into vagina or sterile container at the moment of ejaculation. Fertilization hinges on these polyrhythms, and on the different mobilities of gametes—the sperm to wriggle and 'swim', and the ovum to open and close its protective membrane in an instant when one sperm penetrates it. Fertilization may implicate the cilia present in the fallopian tubes or enroll lumen (biopsy) needles and petri dishes in laboratory settings. It incriminates particular geographies, and their scales and spatialities—among them the arrangements, contours, and configurations of bodies, domiciles, or other spaces of sexual congress, or in vitro fertilization (IVF) and human therapeutic cloning laboratories. Never innocent, fertilization is charged with the routines, variations, motifs, and (e)motions of sexual attraction, or fear, or power, as well as numerous spatial practices and representations that inhere in divergent geopolitics and political economies. Thus, for instance, the fertilized egg in capitalism differs from that found in other systems, and in any such system are to be found shifting geographies of sex and gender politics; of abortion, surrogacy, parturition, parenting, or adoption; of production and social reproduction; of health, illness, and well-being; and of ideology, ethics, and religion.

The term pre-embryo describes a blastocyst or cell-cluster that forms after fertilization and prior to attachment to the uterine wall (England 1983), or location in other media in the laboratory. In what follows, my comments first address the former, uterine, environment, and then the latter, clinical, setting. Whereas it is commonplace to view the time of the pre-embryo as spanning twenty-eight to thirty days, under some schema the period accounts for the first fourteen days after fertilization, during which the blastocyst embeds itself in the uterine wall.[2] The primitive streak appears, and is a key bifurcation process along the embryonic midline; it is deemed a development marking the creation of a unique human being (Bateman Novaes and Salem 1998). In the United States of America, The President's Council on Bioethics (2002, n.p.) has asserted that, at this moment of bifurcation, "the being in question can no longer be anything but a single being" and the possibility of replication is removed. Hence, the President's Council makes further pronouncement that

> there are no sound reasons for treating the early-stage human embryo or cloned human embryo as anything special, or as having moral status greater than human somatic cells in tissue culture. A blastocyst (cloned or not), because it lacks any trace of a nervous system, has no capacity for suffering or conscious experience in any form—the special properties that, in our view, spell the difference between biological tissue and a human life worthy of respect and rights. Additional biological facts suggest that a blastocyst should not be identified with a unique individual person, even if the argument that it lacks sentience is set aside. A single blastocyst may, until the primitive streak is formed at around fourteen days, split into twins; conversely, two blastocysts may fuse to form a single (chimeric) organism. Moreover, most early-stage embryos that are produced naturally (that is, through the union of egg and sperm resulting from sexual intercourse) fail to implant and are therefore wasted or destroyed.

As the primitive streak develops, gastrulation begins—a process marking changes in embryonic cellular structure from one to three layers comprising the endoderm, mesoderm, and ectoderm, and the onset of digestive functions in certain endodermal cells. At this point, the President's Council judges that the embryo exists as a potential child; this on the basis that the rhythms of development now produce nervous, circulatory, and digestive functions, and the possibility of twinning is foreclosed.

Such judgment by the President's Council rests neither on the presumption of the objectivity of scientific data nor on understandings of their patterns and forms, which are contextual. Rather it pivots on debates based in philosophy and theology about whether the form of a life *develops* (epigenesis) or is *given* (preformationism). Adherents to either position hold that, at some point, entering into a conceived entity will be a "driving formal cause, as Aristotle put it, or a vital principle of force, as epigeneticists of the eighteenth century put it" (Maienschein and Robert 2010, 3). Nevertheless, disparity remains between proponents of the two ideas about the timing of the actualization of life in human beings, understood as quickening, ensoulment, or personhood.

Setting aside discussion of this divergence of thought about when personhood may be present in the unborn, and with reference to Streeter's stages, Marjorie England (1983, 14) invokes several kinds of rhythm and movement inside the geography of the pregnant body. Vividly, she describes the 'moments' leading up to the point at which the pre-embryo becomes the embryo:

> the neural groove closes and the primary brain vesicles form. The optic vesicles and lenses form, the otic [auditory] vesicles are present and the brain bends at the midbrain flexure. Limb buds are present. The primordia of liver, pancreas, lungs, thyroid gland, mesonephric tubules and heart appear. The two heart tubes are fused in the mid-line and contractions commence.

England then delineates other stages to delimit the embryonic period to approximately eight weeks' gestation; thereafter, the entity is deemed a fetus. In 'normal' development, change to around week forty is characterized by rapid growth and by the differentiation of organs and tissues formed earlier. Her descriptions are richly spatial, and reference is made to appearance, surface, exteriority, interiority, edge, closure, bulging, invagination/folding, elongation, hemispheres, striation, sinking, division, segmentation, regional differentiation, looping, overhanging borders, hillocks, axes, canals, branching, mid-lines, lateral margins, extension, and movement.

Elemental in England's (1983) work, and Streeter's before her, are the topographies of the embryo and fetus: their dimensions, points, lines, planes, and sections; as well as a layering (up) and stretching (out) over time—from hours to days to weeks and trimesters, and millimeter by millimeter—from crown to rump, potentially growing into a person. Foundational, too, in England's narrative is the uterine territory, abutted by placenta, and bounded by other organs. In a vastly different context, Stuart Elden (2005, 16) has remarked on the manner in which late capitalism "extends the mathematical, calculative understanding of territory to the entire globe". Equally, it is possible to argue that such understanding penetrates the surface and depth of the pregnant body that, as Lianne McTavish (2010) has shown, has been visualized as 'the world'—the cradle of humanity, a site of conquest, and a mappable terrain. Other real and imagined spaces—heterotopia—are implicated in bringing to life, the reproductive laboratory not least among them; their existence underscores another point of Elden's (2005, 16)—that "ontology is not concerned with 'what is', but with how 'what is' is".

As represented by England (1983), Streeter's schema refers to twelve horizons (scripted as I–XII) that typify the 'normal' pre-embryo, and another eleven horizons (XIII–XXIII) that define development in the embryo; thereafter change is marked in terms of the passage of weeks. The use of the term 'horizon' is noteworthy; its etymology intimating that the stages of development are bounding circles, divisions, or separations, and that they are scaled (OED). Thus, pre-embryo, embryo, and fetus can be measured by use of a system of orderly marks at fixed intervals, in millimeters, and their progress calibrated. The scaling of these entities is significant, and the language noteworthy: consider, for example, Horizons IX–X around days twenty to twenty-two when the embryo measures in length 1.5–2.0mmCR (from crown to rump) to around week thirty-eight when the fetus measures 360mmCR. This kind of understanding of fetal development is implicit in and shapes what we know of the world and how we know it. Indeed, and more generally, "scalar stories, frames and metaphors [of which these fetal narratives are a part] . . . operate as yoking mechanisms that name, assemble and delineate sociospatial boundaries and relations—they produce 'world-making' epistemologies and categorizations" (Moore 2008, 214, 221). Such revelations are not simply to be directed to the external world, but operate on, in, and through the reproductive body.

A second setting for pre-embryos and embryos is elaborated by Ariff Bongso and Mark Richards (2004) in work they have done tracing the history of stem cell research, and various perspectives on it. They are especially interested in totipotency in the cells of mammals, noting that the fertilized egg, the zygote that develops from it, and the first two to sixteen blastomeres (cell divisions) that result are examples of totipotent cells. Such cells could possibly give rise to a whole organism, including the whole of the fetus and the placenta, but cannot self-renew and are not, therefore, stem cells. Rather, they note that embryonic stem cells "are derived from the isolated inner cell masses (ICM) of mammalian blastocysts" (829). It is those ICM that will become, at gastrulation, the endoderm, mesoderm, ectoderm, and "finally form the complete soma of the adult organism" (829). In laboratory conditions, embryonic stem cells can be developed into stem cell lines that can be propagated and do not experience senescence because a certain gene, telomerase, "ensures that the telomere ends of the chromosomes are retained at each cell division" (829); this 'immortality' clearly does not arise outside the laboratory. Bongso and Richards also describe the possibilities that exist in relation to fetal stem cells—"cell types in the fetus that eventually develop into the various organs of the body", but note that findings are very limited because of the "unavailability of abortuses" (830). I return to these matters below.

PICTURING LIFE

In 1965, *Life* magazine featured a cover photograph of an eighteen-week-old fetus *in utero* taken by Swedish photojournalist Lennart Nilsson. During the 1940s and 1950s, Nilsson had gained a reputation for taking photographs of war victims and displaced persons, Swedish life and celebrities, and ants and other insects (Conley 2012). However, his particular fascination has been with scientific imaging, epitomized in the depiction of fetuses for *Life*. That work has been based on decades of experimentation with scanning microscopy, endoscopy, and laparoscopy, and on collaboration with medical professionals and imaging inventors and technologists, notably in Germany and Japan (NOVA 1996; Stormer 2008).

The cover image of *Life* magazine and later works by Nilsson serve to politicize how bringing to life is understood (Dubow 2012; Newman 1996; Reagan 2012). Donna Haraway (1997, 27–8) has commented extensively on the manner in which the "visual image of the fetus is like the DNA double helix—not just a signifier of life, but also offered as the thing-in-itself". Of Nilsson's *Life* images she observes that many "were of extrauterine abortuses, beautifully lit and photographed in colour, the visual embodiment of life at its origin. Not seen as abortuses, these gorgeous fetuses . . . signified life itself, in its transcendent essence and immanent embodiment".

Haraway's comments build on her work in *Simians, Cyborgs and Women* (1991), and the influence of her thinking is also evident in two more recent and incisive genealogies of the embryo. Sarah Franklin's (2006, 168) analysis of the embryo as cyborg is deeply spatializing, invoking an "embryo-strewn world" too complicated to allow us to "map the . . . social, political, scientific, medical or ethical lives of human embryos, with all of their increasingly prominent civic and legal entanglements". Likewise, Lynn Morgan (2009) has documented the significance of the embryo and fetus as parts of origin stories and ontologies, examining the ways in which views of them affect understandings of conception, abortion, contraception, IVF, cloning, stem cell research, identity, and nation. A perceptive analysis is undertaken of marketing campaigns that use images of embryos and fetuses, and of websites such as MissPoppy.com, which combines merchandizing with messages such as this: "Protect our troops—from the womb to the war. What if the fetus you were going to abort would grow up to be a soldier bringing democracy to a godless dictatorship?" (in Morgan 2009, 27). Miss Poppy's particular depiction of the centrality of life in the deathly defense of specific (geo)political systems is revealing. Her plea unwittingly underscores the salience of Tamar Mayer's (1999) observations about the gender ironies of nation: that 'the nation' is a feminized construction and fiction of innateness, the property of men, and profoundly fragile—like the thousands of casualties hurt by mobilized, militarized processes of democratization that Miss Poppy fails or refuses to consider. In compelling fashion, then, Morgan (2009) describes figurations of the embryo and fetus as scientific objects with increasing numbers of social—and I would argue spatial—functions. By the start of the twenty-first century, she suggests, the embryo in particular had "escaped the jurisdictional confines of medicine entirely, and found roles in entertainment, art, advertising, legislation, education, commerce, and of course as political propaganda" (10). One of its trajectories has been into a visual economy of the unborn; indeed, the increasing visibility of the pre-embryo, embryo, and fetus arises precisely because it has become possible to trace their rhythms, mobilities, and geographies both inside and outside the confines of the female body (see, for example, Stephens 2010).

Referring specifically to *The Miracle of Life*, Morgan (2009, 14) writes that, even so, the "embryological view of development pretends to exist outside of time". In terms provided by Elizabeth Deeds Ermarth (2011, 89), I think such heterochrony is achieved by "sacrificing what is idiosyncratic for what is common", and what *is* common readily comes to be seen as a universal and naturalized place of origin. However, Morgan offers the apt admonition that "embryos do not carry their meanings intact" (15), thus lending weight to the view that we create the symbolic and material through which diverse cultural, rhythmic, and spatial practices of bringing to life are shaped and circulate. Her arguments are borne out by Sara Dubow (2012, 9), whose historical treatise on 'ourselves unborn' delineates the ways in which fetal life serves as a window into diverse anxieties ranging from race and gender, to parenthood; from religion and ethics to science; and sometimes serving as "a proxy for seemingly unrelated issues like immigration, the Cold War, feminism, or liberalism". In short, fetal life *moves* and configures other geographies. Doubtless then, Nilsson's works form some of the key mechanisms that have unlocked this "figuration of bits of life" (Alaimo 2010, 479), driving their mobilization and complicating what it means to flourish in the world.

Interest in figuration is well established among geographers and others engaged in considerations of space and scale. According to Irit Rogoff (2000, 10), art and visual culture are among the chief means of such figuration, and art is an especially important "interlocutor" in the geopolitics of regions and nations, and in "identity constitution and identity fragmentation" at other scales. Her insights have wider application for thinking about the visual economies, geographies, mobilities, and rhythms of bringing to life. They inform my rereading of Nilsson's science documentary and photojournalism in ways that unsettle geography's complicity in the constitution of subjects, narratives, and languages of signification. Additional aid in interpreting Nilsson's images comes from consummate work on prenatal sublimity and commonplace life by Nathan Stormer (2008). He underscores the importance of making explicit how we look at prenatal images by Nilsson and another visualization expert, Alexander Tsiaras, and reads Nilsson's work as constituting a prenatal sublime focused on "the infinities of the universe" (648), which I think also reveal the spaces of prenatal life to be heterotopia (Figure 2.1).

Michel Foucault (1986, 24) understood heterotopia to function in several ways, not in the least as a mirror, from which standpoint "I discover my absence from the place where I am since I see myself over there". In turn, Stormer (2008, 649) shows how Nilsson creates fertile and pregnant bodies as heterotopian;[3] here are sublime territories whose visualization and

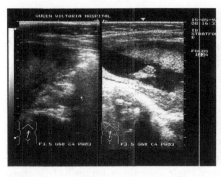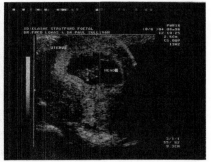

Figure 2.1   Ultrasound heterotopia 1992 and 1996.
Source: Stratford.

mapping remind us that "we are all natives of the womb". Yet at the same time, they are absented from references to the pregnant body and become territories of the unborn "within cosmic and microscopic infinities" (653).

That *The Miracle of Life* is a *moving* picture warrants additional consideration. In general terms, Christina Kennedy and Christopher Lukinbeal (1997) have emphasized the importance of film as constitutive of social knowledge, ideological construction, and hegemony. Ranging from individual stories to comprehensive social, global practices, a film's scalar fluidities are described by them as "heterotopic, functioning on multiple levels of meaning, space, time and geography" (34). Tim Cresswell and Deborah Dixon (2002, 5) extend such understandings to consider the ways in which film is the temporary embodiment of processes that constitute the *mobile* world in terms of the gaze: "Ways of seeing, ways of looking, and ways of being seen," they suggest, "are all still open to question, all constitutive moments in the relational construction of meaning and power". Gillian Rose (2012) has reiterated the ways in which visuality is constitutive of a sense of self, and powerfully affected by film's capacity to structure ways of looking. In turn, Antje Schlottmann and Judith Miggelbrink (2009, 1) have cautioned vigilance among geographers on the basis that "images and visuality could prove to be a blind spot for the very reason that they play such a prominent role in geography". They note the power of images to produce knowledge by means of certain disciplinary and technical conventions: images are a means to access, actively constitute, and comprehend different conditions for the production of the world and spatial understandings of it. Like their still counterparts, moving images rely on metaphor, representation, the emblematic and symbolic, and strategies of localization and personalization. But there is a palpable sense in which film is able to penetrate the interiority of the body, psyche, or territory by means of certain kinds of mobile gaze (Nash 1996; Rose 1993, 2012). Freud wrote of the scopophilic pleasures of the moving image, a matter elaborated on by Laura Mulvey

(1975/1991) in her analysis of visual pleasure and narrative cinema. Mulvey makes the point that the moving cinematic gaze enables objectification and control, including of those unwilling to be observed, or who seem to know nothing of the scrutiny to which they are subjected—and prenatal life typifies the latter. So, the pre-embryo, embryo, and fetus are constituted as both producing and being products of specific rhythmic and moving changes in what is presumed to be the ontologically stable interior of the female body. In this regard, like Carol Stabile (1992) before him, Stormer (2008, 661) is at pains to describe how woman is then made either to disappear from the frame or reduced to a use value. This tactic "exemplifies the trope of unveiling nature wherein a masculinized rational gaze penetrates the feminized body of nature". It also illustrates the multiscalar affect/effect of bringing to life.

READING NILSSON'S DOCUMENTARY

Informed by the foregoing insights, what follows is a detailed reading of *The Miracle of Life*, in which attention has been paid to the rhythms, mobilities, and geographies apparent, implicit, or perceptible in the film, and which aid my reflection on the figuration of bringing to life. The documentary begins with an on-screen printed caveat and overlain narrative, spoken in mildly authoritative tones by a woman with an American accent: the film is about human reproduction, and discretion is advised. Reproduction is presumed natural, heterosexual, and heteronormative. Of such widespread assumptions, Phil Hubbard (2008, 642) has noted that:

> ideas about sexuality often draw sustenance from the biological 'fact' of someone's anatomical gender. Women, it is assumed, are normally sexually attracted to men, and vice versa, with procreation and bringing up children seen as [the] fulfilling outcome of 'sexual congress'. The seemingly indisputable fact of reproduction—an egg meeting a sperm—supports the idea that this is the natural (and perhaps 'god-given' order of things), and that other sexual possibilities are simply aberrations.

The opening shot is Earth's curved horizon, foregrounded by green-grey rocky islets in seas of varying hue—azure, cerulean, and midnight. This is the blue planet four billion years ago, an infant on whose surface swirl masses of cosmic dust and particles, some to be engulfed in shallow and primordial seas; here, too, is a trope of sexual union. Beneath the voice of the narrator, the sounds of rushing wind and wave are discernible. In those seas, "the miracle of life began" (1 min 33 sec); protozoa, "the first organized form of primitive life" (1 min 45 sec), used cilia and flagella to move; from them, we are told, evolved all else. In this initial sequence of the documentary, the living planet Earth is conceived, its spaces colonized in descending scale from the planetary to the cellular. By means of binary fission or division, a single cell produces all life on Earth. DNA[4] molecules are positioned along each of forty-six chromosomes in every cell; they are the only entities capable of identical replication, their chemical composition across species virtually the same. It is their arrangement on chromosomes which determines, first, cellular function and, then, different forms of life—single cells, plants, and animals, including human beings. Life, we are told, depends on these processes of replication and differentiation. Such terms are used directly and obliquely in *Rhythmanalysis* when Lefebvre (2004, 6) remarks:

> In the field of rhythm, certain very broad concepts nonetheless have specificity: let us immediately cite **repetition**. No rhythm without repetition in time and in space, without *reprises*, without returns, in short without **measure** . . . But there is no identical absolute repetition, indefinitely. Whence the relation between repetition and difference . . . there is always something new and unforeseen that introduces itself into the repetitive: difference (original emphasis).

Then, across the screen, a man and woman swim in clear blue water, their presence there alluding to the primal seas and the start of life on Earth. Here, however, the narrator notes that, for humans, "each beginning" (5 min 50 sec) is located in the ovaries. A dramatic cut from the swimming female form moves the gaze from the seas into the interior of the female body, where two hundred and fifty thousand ova are already mature even before her birth. Each oocyte has twenty-three chromosomes, half that present in a 'normal' cell, and something like six hundred will rhythmically be shed over the course of a female's reproductive life. Without the egg, there is no embryo. But we now know that in the patterned development of ovarian tissue are to be found mechanisms that make possible the fertilization of ova harvested from aborted female fetuses (Gosden and Lee 2010); it is, then, feasible to make life from a life that was never born. Of course, *The Miracle of Life* is silent about such possibilities, which (publicly) post-dated the documentary. Nevertheless, such options now exist.

The basic elements of the female's twenty-eight-day reproductive cycle are then described. Under 'normal' circumstances, this hormonal cycle is itself a rhythm during which, at mid-cycle and ovulation, a follicle swells with liquor folliculi to the size of a sand-grain, then ruptures, driving the ovum into the fallopian tube. The fluids encasing the egg have the same salinity as the sea—a reference back to the narrative of evolution with which the documentary started. The engorged fringes of a fallopian tube, known as fimbria, "search for the newly released egg" (9 min 12 sec), brush against the swollen follicle and, as it erupts, sweep the ovum into the deeply creviced tube whose interior is only twice the thickness of a human hair. Imperceptible muscular contractions in the fallopian tube propel the ovum down to the uterus—a five inch (almost thirteen centimeter) distance that will take between three and five days, assisted in that journey by tiny cilia reminiscent of those on a protozoa. If not fertilized within approximately twenty-four hours after leaving the ovary, the ovum will begin to disintegrate and be shed in menstruation, another metrical flow and pattern. The language deployed by the narrator is reminiscent of that invoked by Marjorie England (1983) in her elaboration of Streeter's horizons of prenatal development.

All the while, Nilsson's images scaffold the narrative, writing the space of the body's interior, mapping it, constituting what was then a new visual language, a new geography. Like his photograph from 1965 of the fetus in *Life* magazine, many of Nilsson's moving images in *The Miracle of Life* are dramatically framed in black, the lighting chiaroscuro, the hues predominantly gold through red, or azure to deep blue, the sound track primarily the sound of rushing wind or water, with themed passages of electronic music beneath. Filmed at very high levels of magnification, the images alternate from the interior of the female body to cells, cilia, and passageways and conduits; Stormer (2008, 655) describes this visualization as "vertiginous".

A dominating idea is of the ovum on a 'progress'. Since the early fifteenth century, this term has been understood as walking forward—questing—and in a figurative sense marks growth and development. From time to time, in sharply delineated shifts in Nilsson's imaging, that movement stops, the warm tones of blood-engorged tubes carrying matter replaced by stasis and the blues and purples of stained cell sections illuminating the form of the egg and its interiority. At one point what are said to be crystallized and stained sex hormones linger on the screen; these are strange and kaleidoscopic images. They are also cut loose from any anatomical anchors as they surge, flow, blossom, twist, and spike: metonymy for transformations that hormonal changes precipitate. As the narrator explains how, by seven weeks after fertilization, hormones sensitize the fetus to be female or male, it is instructive to recall the etymology of hormone, from the Greek, also *hormone*—that which sets in motion, shared with the derivations for the words emotion and wave (OED).

A shift in emphasis then propels the viewer onto the streets of a busy city, amid dozens of walking feet, as the narrator describes the world's population in terms of its billions—an estimated seventy-seven billion over the generations. Several minutes of film then attend an examination of the male reproductive system, starting with reference to the four billion sperm produced in the testicles over life. Much is made of the need for the testes to be slightly below core body temperature; hence their encasement and suspension in the scrotum. Again, the narrative is profoundly spatial. Each testicle comprises tightly wrinkled seminiferous tubules which—if rolled out—could be as long as seven hundred feet (two hundred and thirteen meters), and in these about one hundred million sperm are produced every twenty-four hours, alongside testosterone, which is produced in small compartments in the walls of the tubules. As sperm mature, they move closer to the central inner canal of the tubule in which they are produced, and then move again to the epididymis, where they are stored. In both spaces, the sperm are constituted as *themselves* having an infancy that requires the ministrations of much larger and nurturing guardians—the 'nurse' cells. This choice of description is telling: over centuries, the nurse has been figured as feminine, religious, domestic; as handmaiden to masculinized doctors; and as paramilitary, sexualized, or monstrous (Darbyshire 2010). In *The Miracle of Life*, nurse cells are certainly feminized, and their protectively mobile role is described in terms both solicitous and martial, the latter because the interior of the male's body—even as progenitor—remains a potentially lethal environment, the male body sensing the sperm, with only twenty-three chromosomes, as foreign objects to be expelled or neutralized. The nurse cells serve, first, as barriers against white blood cells and, second, as feeding and communication lifelines. One nurse cell protects many sperm, and eventually will 'escort' them to the epididymis, where, for as many as several weeks, the sperm develop the ability to swim. From time to time in this part of the documentary, the invocation of rhythm and movement hints at the possibility of a meta-intelligent driving force.

Also shown in animation are the Cowper's gland, which lubricates, and the spongy prostate gland, which secretes an alkaline fluid to protect sperm—these two provide the chief constituents of semen. The series of rhythmic contractions that constitute ejaculation propel this mix from the body via the vas deferens and seminal vesicle, and the urethra, during which time a sphincter shuts off the bladder. An ejaculation may contain two hundred and fifty million sperm. If sperm are not thrust out of the male body in this way, they will eventually die in the epididymis and be reabsorbed into the body. Several filaments covered in tiny hooks make up the propulsion mechanism of the sperm's tail, and are shown in detail as sperm swim across the screen or appear in vermillion shots of dissected cross-section, showing layers of enzymes and enzyme inhibitors, DNA, and so-called 'fuel packets' to energize locomotion. Again, the descriptions are redolent of diverse rhythms, and of mobile bodies and objects; with the images, the narration writes the geography of the reproductive male body, its inner spaces, and their differential scales. Some twenty percent of sperm are deformed or imperfect—and here, another jump occurs from the miniature world of the interior of the testes to exterior sites of risk and harm to them. Scenes of modern urban life appear on-screen: life dominated by overcrowding, stress, pollution, occupational hazards, radiation, poor nutrition, and constrictive clothing. But the resilience of the (male of the) species is underscored, the cycle of life described as one "fuelled by the need and the drive to reproduce" (26 min 32 sec); this assertion is made as the image shifts to seas crashing onto cliffs, inscribing again the scalar shift between the cosmic and microscopic, and succumbing to a much used trope that invokes climax.

It is in the ensuing discussion on attraction that the full force of the heteronormative assumptions underpinning the representation of human reproduction are apparent. Here, too, rhythm and movement are paramount.

A woman and man are filmed dancing in silhouette to a lyrical theme repeated whenever the narrative addresses social interactions between heterosexual couples. The lexicon includes desire, intimacy, tenderness, romantic love, and affection. Reference is made to "the dance of courtship, which may lead to conception" (26 min 57 sec), and to actions compelled by the "complex rituals of mating" (27 min 18 sec). These rituals are signified by the brain triggering a series of arousal responses: the rhythms and movements of dilating pupils, swelling sensory receptors in the skin, and increasing heart and respiration rates. In a manner reminiscent of a shadow play, a thermal camera traces the upward movement and interior of the penis. Then, to a more strident and staccato musical theme reminiscent of reveille, the gaze is redirected to the interior of the male body. There, large numbers of sperm are seen to assemble, the underlying narrative tracking increasing arousal, punctuated by a huge rush of coordinated movement involving Cowper's gland, prostate, seminal vesicles, and epididymis.[5] Via rapid contractions, approximately half a teaspoon of semen moves to the vas deferens, along some twelve inches (thirty centimeters) of tube, in mere seconds, and into the vagina, at which point there is a reprise of the lyrical dance music first played for the dancing silhouetted couple. This mixing of military precision and romance is uncanny—a juxtaposition of urgent organization, climax, and release.

The narrator then describes how, in the first twenty minutes after ejaculation, the semen slows down and coagulates—perhaps to optimize its chances to stay in the vagina, and to afford to the sperm protection from the environment into which it has been transported. Nevertheless, nearly twenty-five percent of them will die almost immediately in the more acidic conditions. Thereafter, the semen becomes less viscous, and sperm begin to move rapidly upstream against downward currents in the female body. A sense of urgency is invoked in the narration and the sound recording of active sperm, reminiscent of swirling winds and waves first heard at the start of the film in comments about the origins of life on Earth. Viable for less than two days, the sperm are described as having to fight against the hostile environment of the female body—especially its white blood cells and acidity—designed to protect her from invaders. Some sperm get 'lost' and some attempt to fertilize normal body cells. Those reaching the cervix swim up the uterus to the entrances to the fallopian tubes, aided—during ovulation—by conduits of mucin, a kind of protein less than one-one hundredth of a millimeter thick (0.0004 of an inch). Of the remaining sperm, estimated at less than half of the original, only half may swim up the fallopian tube that holds the ovum, and will have to swim against a current produced by cilia in the tube. Some get stuck, others lose direction. At the same time, the enzyme barriers covering the sperm heads are slowly worn away, and they become capable of fertilizing the egg if they encounter it—and only about fifty of the original two hundred and fifty million will do so. Image after image shows the 'battling' spermatozoa on their quest.

Then, when it seems as though all the sperm will be vanquished, the ovum appears, filmed in darkened shades of teal reminiscent of the blue planet. The narrator describes again how the ovum has two layers of nutritive cells and explains how the sperm find it, surround it, and release digestive enzymes to break through these layers; the first to do so is drawn within and, in an instant, will chemically shut all others out. At this point, the narrator notes how "the joint force of their exertions starts the egg rolling around like a mysterious celestial body" (41 min 24 sec), and I think it noteworthy that, in other of his more artistically rendered photographic work, Nilsson has represented egg and sperm as if the former were, indeed, a celestial body, the latter a greatly enlarged seed growing from the earth towards it (Figure 2.2).

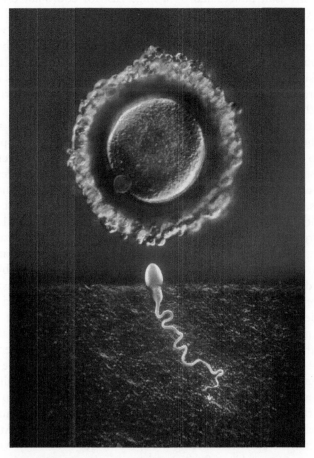

Figure 2.2   A sperm in front of the egg, 2003, computer manipulated image, 3845 × 5562 pixels. Photograph courtesy of Lennart Nilsson/TT [Tidningarnas Telegrambyrå Nyhetsbyrån]. Reproduced with permission.

At this point in the film, still red shots of the penetrating sperm at great magnification are juxtaposed against moving images of the sperm inside the blue egg, its siblings surrounding it and moving rapidly, but without effect; their progress halted. Other shots show the fertilizing sperm's tail dislodge, the head swell and rupture, and strands of DNA material move into the egg, as if "dispersed by an explosive force" (44 min 6 sec). Juxtaposed against these stills are blue images of cell division in a new embryo which has "never before been filmed" (44 min 49 sec). Two nuclei become apparent, and then two cells, and then four, and eight, and a zygote forms on-screen, with division continuing and accelerating. All the while, the zygote moves toward the uterus as a dense and compact cluster and, after five days, it is a blastocyst no larger than the original ovum. Within ten days, the blastocyst implants in the uterus, nourished first by the lining and then by the placenta. Attention is devoted to the kinds of developmental details established by Streeter and noted earlier. The temporal scale accelerates; descriptions of key stages and movements are provided with less detail and more quickly than typified the narrative before fertilization, and embryonic and fetal size are noted for weeks two, four, five, six, seven, ten, eleven, twelve, fourteen, fifteen, and eighteen.

No mention is made of amniocentesis or preimplantation genetic diagnosis, two tests to determine the 'health' of a potential child (Rothman 1986; Thompson 2010). These may, of course, influence decisions to terminate a pregnancy deemed 'marginal' or to *not* implant an embryo using IVF procedures. Once such decisions are made, other pathways, mobilities, and geographies ensue, among them disposal or donation (see Gurmankin, Sisti, and Caplan 2004; Scully, Rehmann-Sutter, and Porz 2010; Svendsen and Koch 2008). No mention is made, either, of the critical importance of location in determining the legal status of the embryo or fetus (on which, see Annas 2007). In this regard, there is significant debate about the differential status of embryos inside the uterus and those used for medical research, the location of which is in laboratory petri dishes. In these very terms, Radhika Rao (2010, 39–40) refers to the manner in which location was dismissed as salient in one case heard in the United States of America. Specifically, in *Kass v. Kass* (1998):

> the trial court refused to consider location, granting a woman the same privacy rights over in vitro embryos that she would possess over in vivo embryos. In that case, the court awarded five frozen embryos to the genetic mother to be implanted and carried to term over the objections of her ex-husband, the genetic father . . . this on the basis that in vitro status should give no additional rights to the father because it doesn't matter if fertilization took place 'in the private darkness of a fallopian tube or the public glare of a petri dish'.

In the final analysis, the realpolitik of bringing to life is absented from *The Miracle of Life*. Instead, the final moments of the film are footage of developmental stages prior to the eighteenth week, including both live embryos and fetuses and, as Haraway (1997) reveals, abortuses. The last of these images is a live fetus, moving in amniotic fluid to a rushing, rhythmic sound of fetal respiration. The narrator falls silent, the lyrical music that signifies romantic love and human relations returns a fourth time, and the final, graphic birth scene shows a Caucasian heterosexual couple and female attendants at a hospital delivery of an apparently 'normal' and healthy girl.

OTHER MODALITIES OF BRINGING TO LIFE

Not surprisingly, Nilsson's fetal images and accompanying narratives invoke diverse challenges, some of which have been noted above—such as the option to use preimplantation genetic diagnosis. Much has been written about his work, among them straightforward synopses (Conley 2012) and more complex discussions on figuration or aspects of the visual (Stormer 2008); critical evaluations of the constitution of fetal personhood (Newman 1996); and the images' entanglement in discourses as diverse as those promulgated by the pro-life movement or those working on somatic cell nuclear transfer experiments, otherwise known as SCNT or cloning (Maienschein and Robert 2010). In turn, Sarah Jain's (1998) probing examination of an article in *National Geographic* about Nilsson's work positions his images as geographical representations: the body a wild inner space, a frontier for exploration. Jain argues "that this legibility is embedded in a set of liberal individualist assumptions and discourses that allow for the *pictured body fragments to stand in for all bodies*" (n.p.; emphasis added).

If body fragments can perform such functions, then might they also stand for all life and, if so, what other modalities of bringing to life are invoked here, and what implications have they for our conduct and capacity to flourish? How, in short order, to account for the rhythms, mobilities, and geographies of, for example, IVF or parthenogenesis—reproduction in which the growth and development of embryos occur without fertilization? One way into this domain is by reference to a masterful examination of *the politics of life itself*, in which Nikolas Rose (2007) considers developments that enable bringing to life in the biotechnological century, his work returning us to questions of conduct and flourishing. Rose seeks to produce a 'cartography of the present' capable of destabilizing and reshaping the future: how do we want to be in the world in light of new biotechnological advances, and who should decide, for example, such matters as embryo selection or the designation of life or its termination? Referring to the governmentality of vital existence at the molecular scale, Rose ponders the emergence of knowledge deeply vested with emotional energy, new territories for bioeconomic exploitation, new ethics, and new organizational principles for their understanding. Part of his focus is on how visualization is key to the discourses of bringing to life, itself framed in the terms of a "style of thought . . . ways of thinking, seeing, practicing only possible within that style—membership, practical arrangements, institutions, modes of questioning, criticism, seeking errors and corrections" (12). We are, then, bearing witness to a new "molecular vital politics" (4) in which, as Rose describes it, life is decomposed, anatomized, manipulated, amplified, and reproduced; DNA is bound; and chromosomal structures are made visible, separated, or cloned.

Such possibilities apply to more than just human life as well; they implicates the rhythms, mobilities, and geographies of other species. This interspecies dimension of the issue concerns, for example, Jose Cibelli and Kai Wang (2010) in their analysis of decisions made in 2007 and 2008 by the United Kingdom's Human Fertilisation and Embryology Authority (HFEA). It first determined that embryos created from human DNA and the oocytes of other animals should be deemed human and subject to its regulation; and then approved two research proposals to manufacture human/non-human animal hybrid embryos. Focused on the wider notion of future human perfectibility across generations, Peter Scott and Celia Deane-Drummond (2006, 4) further underscore the complexity of the debate aired by Nikolas Rose (2007) about the creation of hybrid embryos when they observe that as

> human identity raises the matter of human nature so the issue of technological practices in medicine raises the matter of the situation of technological change. That is, in order to answer the question as to how the *field of moral action* is structured in this new and contested area, we are caught up in the politics of technology, including the politics of the meanings of technology. In other words, technological practices are always present through *networks of power*. (emphasis added)

In my estimation, these politics are necessarily a kind of *geopolitics*, and Scott and Deane-Drummond's commentary illustrates the debate's capacity to move and constitute new relational geographies. Indeed, Rose (2007) makes the case for how these vital politics are mobile at multiple scales—at the molecular level, for instance, or in terms of prudent citizenship. For him, this vital prudentialism is entangled in "an ethic in which the maximization of lifestyle, potential, health, and quality of life has become almost obligatory" (25) and it is deeply implicated in questions of conduct in the present and for the future. These particular insights are not new; what *has* changed are the ways in which these mobile "circuits of vitality" (38) have been disembedded—displaced. Thus, prenatal life is central to Rose's argument; at issue is how vitality has been

> decomposed into a series of distinct and discrete objects, that can be stabilized, frozen, banked, stored, accumulated, exchanged, traded across time, across space, across organs and species, across diverse contexts

and enterprises, in the service of bioeconomic objectives . . . [a matter that] raises questions about the borders of life, and those troubling entities—notably embryos and stem cells—whose position on the binaries of life/nonlife and human/nonhuman is subject to dispute. (38)

Developing this argument, Rose formulates the ways in which life is reframed as information, especially with the mobilization of DNA, its modelling, simulation, and movement. Ultimately, however, there is knowing life as information and knowing life; this distinction, Rose writes, "must be found in the life-liness of the created entities themselves" (48). Among them are embryos in diverse embodied or disembodied locations; embryos in variegated states of belonging (with those deemed spare and surplus now having official and legal status); and liminal entities such as sperm, ova, or blastocysts. Then there are other, different, and equally intensely debated entities that bring into sharp focus the "very meaning and limits of life itself" (49): stem cells, stem cell lines, and embryoid bodies that have reached gastrulation and will produce differentiated tissues or specific enzymes. Rose is clear about the geographical importance of these debates: they are sites in advanced liberal democracies that mesh biocapitalism, biotechnology, genetics, risk, and the prudential citizen. Through such debates, life is constituted as "a project, framed in terms of the values of autonomy, self-actualization, prudence, responsibility, and choice" (125).

In very different ways, the moral geographies of this new molecular vitality also concern the ethicist Michael Northcott (2006). In an analysis of the ethical dimensions of human therapeutic cloning, he begins by referencing the first license granted in the United Kingdom to undertake such work. Northcott (2006) argues that therapeutic cloning may imperil human dignity and integrity on the grounds that to experiment with an embryo is to experiment with an entity that has the potential to become a person. He quotes Suzi Leather, the Chair of the Human Fertilisation and Embryology Authority (HFEA), who describes the purpose to permit human embryos to be created "by inserting the nuclei from human skin or stem cells into human eggs . . . to increase knowledge about the development of embryos and enable this knowledge to be applied in developing treatments for serious disease" (quoted at p.73). The procedure being referred to is parthenogenesis, which Paul De Sousa (2010) describes as unlocking the developmental potential of the ovum without the need for sperm to create embryos. De Sousa emphasizes the point that this possibility may unsettle the special status assigned to the embryo, which was recognized by The President's Council on Bioethics when it invoked the 'potential child'. De Sousa's point is that, along with such technologies as SCNT, parthenogenesis enables embryos to be constructed for biomedical purposes and by means other than fertilization, "although the viability of these forms of embryos is comparatively limited" (71). Indeed, De Sousa constitutes a new definition for the embryo so designed: "an entity composed of one or more cells with the potential to form a physiologically independent organism provided it is enabled to do so . . . [The definition] would recognize that in the absence of a reproductive intent, potential embryos are no different than any other tissue" (73). For De Sousa, this reinscription focuses attention away from the dilemmas of embryonic personhood, a matter of no small significance given growing evidence of the totipotency and pluripotency of cells at various stages of development—their capacity to become many forms and serve multiple functions. Instead, donor consent and the provision of adequate information would be "the primary ethical and legal considerations associated with the use of eggs and potential embryos for research or translational applications" (73). But Northcott's (2006) concerns extend beyond the designer embryo, his geographical labors underscoring the proliferation of mobilities and rhythms involved in bringing to life; the divergent understandings of what that phrase might now signify; and what he sees as the moral dilemmas attending what to do with surplus and spare eggs, or those damaged by processes involved in IVF (see also McLeod and Baylis 2010; White and Bluhm 2010). Thus, Northcott returns to the problematics of shifting places of origin—a concern which opened this chapter. In making the case that embryos are ensouled, and that those *"which are not in wombs are in the wrong place"* (82, original emphasis), Northcott views as morally wrong the occlusion/exclusion of oocytes and embryos from "their natural and social environment in the human body and in kinship relationships . . . [such that they] consequently become artefacts, objects of human making and experimentation" (74). Essential to his position is a critique of the ways in which embryos come to be morally and socially 'disembedded' from the human (female) body and from social relations. For Northcott, this process is a *removal*, a displacement that, by means of distancing, "facilitates their use as means to a potential therapeutic end" (76). Such movement, he suggests, is at the heart of a "new geography of embryos . . . [one] of exile" (82).

I began this chapter by reference to Nash's (2005) call for more work by geographers on our shifting places of origin. Here are diverse and immensely potent rhythmic lineaments of everyday life that are heavily weighted with power, as Edensor (2010b) would have it. Here are particular intervals—openings, spaces—by which to consider those vital conjunctures to which Johnson-Hanks (2002, 867) gestured when she noted that most "vital events . . . [are] negotiable and contested, fraught with uncertainty, innovation, and ambivalence . . . The fact that vital life events are rarely coherent, clear in direction, or fixed in outcome dramatically limits the usefulness of the life cycle model". There are manifold politics of mobility here—rhythms, movements, scales, emplacements, and spatial relations of bringing to 'life', whatever that may mean: blastocysts in suspended animation or traversing fallopian tubes and embedding and growing large in uterine walls. There are embryos entangled in international transport logistics; fetuses floating in glass containers on the shelves of medical collections; and live births and still.

In closing without seeking or knowing how to resolve these conundrums, let me return to Bongso and Richards (2004, 828), who make the following observations in their paper on the history of, and perspectives on, stem cell research. First, in terms of bringing to life, they acknowledge that stem cell research and stem cell therapies are wrapped up in "hype and media frenzy" as well as numerous and varied "political agendas and numerous religious and genuine ethical concerns". Second, they suggest that the impassioned tenor of debate arises, at least in part, because stem cell therapy is held up as an immensely powerful—one might even say totipotent—possibility by which to "affect the lives of millions of people around the world for the better". Finally, they note that fulfilling this possibility requires overcoming "numerous technical, legislative, ethical and safety issues". In this observation lies the dilemma of how to flourish, and how to conduct ourselves.

NOTES

1. *The Miracle of Life* was directed and produced by Bo G. Erikson and Carl O. Lofman. The original soundtrack mentioned in this section was composed by Anders Berglund from Metronome Studio. The work was narrated by Anita Sangiolo. Copyright is owned by the Swedish Television Corp; and for NOVA, writing and production were done by Bebe Nixon. The year the film was released, it received an international Emmy and a George Foster Peabody Award, the latter bestowed for "distinguished achievement and meritorious service by broadcasters, cable and Webcasters, producing organizations, and individuals" (The University of Georgia 2012, n.p.). The following year, the documentary won several other awards, and is claimed to have "educated a whole generation of students, making an impact on the face of society as this generation matured" (Maayan 2012, n.p.).
2. Ectopic pregnancies evince an exception to this general rule, and underscore how the rhythms and mobilities of conception sometimes involve various forms of displacement.
3. The instantiation of transgender pregnancies exemplifies this idea of the heterotopian quality of prenatal life and its power to reconstitute significations and meanings of sex and gender; the capacity to hold on to the idea of a normatively sexed pregnant body is profoundly unsettled in the process.
4. Rosalind Franklin's x-ray diffraction images, taken in the early 1950s, revealed DNA to be patterned rhythmically in the form of a simple and stable double helix, describable in terms of axes, horizons, and equators (Franklin and Gosling 1953). This double helix encodes the genetic instructions that give rise to organisms; its structural and functional significance understated by Watson and Crick (1953, 737) when they wrote that it had not "escaped our notice that the specific pairing we have postulated immediately suggests a possible copying mechanism for the genetic material".
5. At this point, anyone who has viewed Woody Allen's (1973) film *Everything You Always Wanted to Know About Sex But Were Afraid to Ask* is likely to recall the seventh of seven vignettes, entitled 'What happens during ejaculation?'. In that scene, the interiority of the male brain is likened to a mission control center, the sperm to military or paramilitary troops, and Allen, dressed all in white, is being coached by another sperm in order to overcome his growing anxiety about his pending 'mission'.

3   Fluid Terrain

In the National Gallery of Scotland in Edinburgh is a large work by William McTaggart entitled *The Storm*, in which a sense of the monumental power of the elements is unleashed. Completed in the studio in 1890, *The Storm* was based on a smaller piece executed in 1883 when McTaggart visited Carradale village on the east coast of the Mull of Kintyre, the peninsular region of his birth (National Galleries Scotland n.d.). The utility of McTaggart's original *plein air* study is evident in the light quality that characterizes the finished studio work. As Tim Ingold (2005, 97) argues, "light is fundamentally an experience of being in the world that is ontologically prior to the sight of things. Though we do not see light, we do see in light". McTaggart's image exemplifies this observation; when one actually stands in front of the work it *feels* illumined.

The painting is a powerfully rendered geography (Figure 3.1). Black clouds roil over churning whitecaps that surge onto unforgiving rock, beyond which are the variegated neutral colors of dunes and pastures. One can imagine into the work the evolution of shorelines and assorted geomorphic processes at the coast and ponder how—eventually—even the rock will succumb. Then there are the varied biogeographical relations between plants, animals, and people that produce this particular landscape which, in time, will also change. Implicit in the scene are the geographies of the islands and highlands that fuel diverse narratives about the resourcefulness and resilience of rural, coastal, and island peoples, and that might flow out to other spaces and places. In fanciful moments, one might even imagine in McTaggart's work the primordial seas that occupied Nilsson, Erikson, and Lofman (1982/1986) in their figuration of the 'miracle of life', as noted in chapter two.

The Storm is also full of complex mobilities and rhythms. Beaches are awash with breaking waves that promise to wreak havoc on fishing nets suspended on tall poles in front of small cottages perched near the seashore. Close study of the original painting reveals a fishing boat precipitously close to disaster just outside the surf, its sails in taut disarray. Villagers gather at a stretch of sandy beach to launch a rowboat whose crew is intent upon rescuing their fellows; behind them, other boats are embedded in taller dunes. There is the agitation of weather and tide, the locomotion of wooden oars,

Figure 3.1 William McTaggart (1835–1910) *The Storm*, 1890, oil on canvas, 122cm × 183cm. Courtesy of the Scottish National Gallery. Reproduced with permission.

and the purposeful movements of people who would seek to deny the waves any spoils. This is a scene filled with earth and sky, men and women, children and the elderly, and crofters and fishers—all manner of mobilities and moorings, as Adey (2010b) might call them. Implied in the scene are ventures in near-shore and deep-water fishing by which vast expanses of coast and ocean have been traversed; trips to nearby villages or further afield; regular and irregular traipsing from one cottage to another, to boat or garden, church or school.

These geographies, mobilities, and rhythms underpin a further powerful element of *The Storm*: its capacity to draw upon narratives about our shifting relationships with sea, sky, and land. Consider the ways in which the patterns of wave-fall and wind-flow differ from those manifest on calmer days; or the storm's capacity to disrupt natural and social patterns and processes, replacing them with others that have their own meter—surge, break, crash, suck, surge. Consider, too, the instance and intensity of the ferocious weather, as well as its duration, which Lefebvre (2004) refers to as birth, growth, peak, decline, and end. And, finally, as the gaze settles, note how detail in the foreground emerges: at the rise of the tallest dune on the bottom left of the painting, and presently out of harm's way, two *children* sit—one seeming to gaze back at the viewer. Others lie on their stomachs, looking out over a small group of adults gathered below. Here is a still corner, perhaps of anxious waiting, an interval and countermeasure to the scene unfolding at the shore.

The Storm is intrinsically interesting, but in this chapter its primary utility is to be emblematic of the geographies, mobilities, and rhythms of climate change, a matter of particular salience for islands and islanders—and especially their younger residents, who concern me here. Already there is a growing body of persuasive writing on climate change impacts upon island nation states and subnational island jurisdictions (Barnett 2012; Barnett and Campbell 2010; Betzold, Castro, and Weiler 2011; Farbotko 2012; Finin 2002; Mimura et al. 2007; Mortreux and Barnett 2009; Roy and Connell 1991; Tompkins 2005; Veitayaki 2010; Walker 2012). At least some of that writing is highly critical of discourses that tend to represent islands as instrumental symbols of global loss and topographical equivalents of a red list of endangered species, their governments and peoples doomed mendicants. At the same time, scholars agree on the need to consider the marked challenges posed by the prospect that entire national populations may be left homeless and stateless as a result of climate change (Badrinarayana 2010a, b; Burkett 2011; Kelley 2011; McAdam 2010; Rayfuse and Crawford 2011; Yamamoto and Esteban 2010). To such ends, significant efforts are being made to think through the implications that might attend novel forms of sovereignty that could account for loss of territory from inundation and other effects of climate change.

Until recently, much writing on such matters has made reference to children and young people primarily in the abstract, albeit important, terms afforded by discussions of intergenerational harm, rights, responsibilities, ethics, and impacts (Gardiner 2010b; Jamieson 1992; McAdam and Saul 2010; Traxler 2002). Lately, however, several shared conclusions about the effects of climate change have been drawn in relation to children's increased exposure to risk, threat, danger of malnutrition, loss of habitat and land, economic hardship, displacement, disaster, morbidity, combinations of cognitive, affective, and physical health issues, and general and pronounced insecurity (Burton, Mustelin, and Urich 2011; McMichael and Lindgren 2011; Sheffield and Landrigan 2011; Tillett 2011; Urbano et al. 2010).

In light of the foregoing discussion, recalling the infant Robin Walker from the *0 to 100 Project* introduced in chapter one, and thinking about children and their place in the life-course, I use McTaggart's work here as the catalyst to ask *how might one think about the geographies, mobilities and rhythms of anthropogenic climate change in relation to young islanders and questions of citizenship?* Assuredly, artists are implicated in shaping varied geographical imaginaries around climate change and, in this sense, are important interlocutors in social and cultural life, often influencing others' spatial understandings in the process (Rogoff 2000). Moreover, artists can be significant role models for young people (Buckley et al. 2007; Hall, Thomson and Russell 2007) and, where they work in political modalities, may influence and be influenced by the sense that children have of their own capacities for civic engagement (I am not, at this point, passing judgment on the merits of any given political messages that may be conveyed). Either way, as Tracey Skelton (2010, 146) has noted, "young people are political actors now; they are not political subjects 'in-waiting'". In relation to climate change, then, enabling children's political engagement is critically important. Whereas this point should stand irrespective of the numbers involved, it is, I think, noteworthy that over one and a half billion people worldwide are aged between ten and twenty-four years of age, and seventy percent of them are in developing countries. In this respect, "young people will be dealing with the threats and opportunities of climate change whether they choose to do so or are forced to do so, and whether they like it or not" (Caparros, Laski, and Bernhardtz 2009, iv).

CONTEXTUALIZING THE GEOGRAPHIES OF THE CHILD CITIZEN

Biologically, childhood marks the interval between infancy and puberty—terms that mark, at one end of a continuum, an inability to speak words and, at the other end, a level of maturity sufficient to take on the responsibilities of adulthood. Between times, there is much focus on conduct: in the *0 to 100 Project*, for example, AJ Spatola, aged eight, referred to "getting into trouble. Yeah, that's pretty, you know, bad". Meggan Scott, at eleven, revealed that "I get off with a lot of things that I do wrong". It seems that conduct is central to their concerns. In this vein, Article 1 of the United Nations Convention on the Rights of the Child defines childhood as the period prior to eighteen years of age, except where national laws deem adulthood to be granted earlier or later (United Nations. Office of the High Commissioner for Human Rights 1989). Subnational jurisdictions may add other guidelines about the rights afforded to minors in order to establish what forms of conduct—such as the age of consent—are sanctioned prior to reaching legal majority or adulthood (Australian Government. Australian Institute of Family Studies 2013).

There are other understandings of this interval in the life-course. Where, for example, much of the literature on the epidemiology of climate change is based on broadly biological understandings of childhood as comprising several contiguous, stable, and natural developmental stages, social and cultural studies of climate change impacts often rest on constructivist perspectives. Significant tensions typify these understandings of childhood and, according to Nick Lee and Johanna Motzkau (2011, 7), they also illustrate the existence of a questionable assumption "that the social and the biological are ontologically separate spheres of activity governed by laws and processes that are, for the most, part incommensurable". Lee and Motzkau assert a pressing need to rethink biosocial dualisms that characterize debates in children's research, drawing on Alan Prout's (2005) idea that hybridity could better frame research engaging children and childhood. Lee and Motzkau present a new navigational guide they hope better suited to contemporary biopolitical understandings of childhood, which is of interest here given my own concerns with questions of conduct and flourishing. The authors justify their task on two grounds: there is a growing appreciation of the socio-natural and socio-technological dimensions of life's processes, and there is need to ask how to ensure that social innovation keeps pace with changes in the environment and climate. In constituting this new guide, Lee and Motzkau draw on three ideas about *life, resource,* and *voice* indebted to Deleuze and Guattari (1988), and to Foucault (2007, 2008). They suggest, for example, that in the modern state—a geographical idea that also concerns the development of populations—children are specifically understood as 'human futures' or 'human becomings' in ways that continue to undergird biosocial conceptualizations of them (Malone 2012; Rose 2007). Diverse strategies and tactics thus come into play to intervene in children's *life* processes, ostensibly—and depending on cultural values—to enable forms of conduct through which they might (minimally) function or (optimally) flourish, both as social *resources* and resourcefully. This intervention is both biopolitical and governmental. Several dynamics thus mark the "circumstances in which children can and cannot find *voice,* along with the range of institutional and technological conditions in which their voices are interpreted, mediated and amplified" (Lee and Motzkau 2011, 11). Who can speak for whom, who is absented or rendered silent? These abiding concerns have motivated efforts to redress the marginal place of children in much of geographical research.

In work framing an agenda for children's geographies, Sarah James (1990) has suggested that geographers (and others) forget to examine, or otherwise render absent or silent, the socio-spatial relations of childhood. In part they err by assuming that children mirror adults and can be enfolded in the work done with older populations; and in part they err by being reluctant to deal with the challenging logistical, scholarly, and ethical dimensions of research with children. James reasons that it is critically important to overcome these errors for two reasons that initially seem rather instrumental. First, she notes that working with children is motivated by the idea that we learn about ourselves by reference to what we were and from whence we came. Second, she suggests that adults learn about the everyday, microcosmic, and immediate by working with children—which, *prima facie,* may not do justice to young people who undergo large-scale forced or freely-experienced travel, or who may have much to reveal about other scales of understanding. However, James's analysis then focuses on three other reasons for undertaking geographical research with children that do emphasize its intrinsic worth *to young people.* Such research may reveal the socio-spatial relationships of their lives, among them those linked to land-use decisions, environmental threats, and children's economic and political dependence on others. It may illuminate diverse forms of spatial behavior, in both public and private spheres, in relation to play, in interactions with adults, in engagement in work, and so on. It may also afford opportunities to describe and understand children's different patterns and processes of environmental cognition. James then outlines a range of possible approaches to geographical studies of and with children, favoring those in which children are *not* inevitably constituted as distinct subgroups of larger populations. Instead, she calls for analyses unsettling the "nature and basis of children's subordinate position and its links with the environment and with geography" and for studies contributing to wider understandings of other variables—such as class, gender, ethnicity—that interact with "age to produce complex patterns of dominance and subordination" (282).

The agenda James proposes and others like it have been taken up by numerous geographers and is reflected in the journal *Children's Geographies* (Matthews 2003). Part of the emergent debate on children's geographies centers on methodological parameters (Hopkins and Bell 2008; Horton, Kraftl, and Tucker 2008; Kesby 2007; Porter, Townsend, and Hampshire 2012). Part of the conversation involves consideration of ethics. For example, research that has utility only to investigators, disciplines, or organizations is (properly) deemed insufficient justification for engagement with children (Ansell 2009; Skelton 2010).

In this vein, Nancy Bell (2008, 8) describes how she had cause to ponder "what a rights-based approach to conducting child research entailed . . . [and to argue] that it is critical for research ethics guidelines to reflect human rights principles that also incorporate special considerations reflected within children's rights instruments". Key among those instruments is the aforementioned United Nations Convention on the Rights of the Child, which establishes that children are moral agents. Article 13(1) of the Convention notes: "The child shall have the right to freedom of expression; this right shall include the freedom to seek, receive and impart information and ideas of all kinds, regardless of frontiers, either orally, in writing or in print, in the form of art, or through any other media of the child's choice" (United Nations. Office of the High Commissioner for Human Rights 1989). One of the bases for Bell's questions has been the growth in number of participatory research projects with children. Published just months earlier than Bell's work, an analysis by Fionagh Thomson (2007) of methodologies for children's research is noteworthy in seeking to unsettle certain elements of such participatory approaches. Thomson's twofold response to standardizing or homogenizing tendencies in research practice is, first, to assert the need to remain vigilant in refusing to view identity as fixed and, second, to call for fewer checklists about how to manage children in research engagement. Instead, she recommends better methods to manage ourselves as researchers, and looks forward to being part of a larger effort that clears "space for more local research narratives to allow space for disagreement and discussion" (216). In my estimation, this goal informs how we might think about the challenges that will confront young islanders in times to come, and is constitutive of a wider and critically important debate about children and citizenship.

Silvia Golombek (2006) refers to four notions of citizenship: *jus solis*—signifying individuals born in a particular country; *jus sanguinis*—marking the children of parents born in that country; naturalization—the acquisition of citizenship by choice, where offered; and reaching one's majority. All such forms of citizenship are attended by rights and responsibilities that imply particular modes of conduct but none, Golombek argues, assumes that children are actively engaged in life *in the capacity of* citizens (see also Hart 1997; Jans 2004). Rather, in programs that include employment, sports, education, school governance, or voluntary labors, children's engagements are seen as experiences through which children learn and prepare for adult citizenship. Such views, according to Golombek undermine "the active status of children in constructing and determining their social lives, the lives of others, and their surroundings" (14). These views also perpetuate a sense that the members of this cohort represent problems to be solved. In short, in terms supplied earlier by Gill Valentine (1996), the social construction of children as human becomings warrants critical engagement in order to *recognize* the "important contributions to civic life by the youngest residents" (Golombek 2006, 28).

Recognition requires deep cultural change, and always needs to be spatial: to consider space, place, scale, and environment; mobility, motility, and stasis; and the patterns, intervals, or rhythms that affect children's varied lives. These are central concerns for Anne Graham and Robyn Fitzgerald (2010) in work that asks how to think about children's participation, interpret that engagement, and understand how it is practiced. Graham and Fitzgerald suggest a trilateral framework to appreciate the importance of an extended notion of children's citizenship: first, there are things to learn from children (enlightenment); second, there are capacities and competencies possessed by children that they should be enabled to exercise (empowerment); and, third, there are virtues to fostering their participation in social life given they are 'natural' citizens who will be formally politically enfranchised on assuming adulthood (what I have called emergence in lieu of a third term nominated by the authors). Several insights from Graham and Fitzgerald's work have wider salience, not least the understanding that children seek to be treated respectfully and as separate from their caregivers, and want participation in social life that is genuine, based on sharing, able to inform their decision-making, pitched at their developmental level, supportive without being cloying or patronizing, and asking them to assume only responsibilities for which they are fully prepared. For Graham and Fitzgerald, these provisions are optimally secured relationally and dialogically, and require that children are recognized as agents in the first place. The authors stress the importance of dialogue and conversation—note that the etymology of these terms resides in notions of dwelling, flows of meanings with self and other, and change. They then delineate a range of ontological, epistemic, and ethical implications of adopting a dialogic understanding of children's citizenship. Ontologically, conversation with child citizens requires dissociating from constant reliance on monological communication flows *from* adult *to* child. Epistemically, such conversation means taking "the self-understanding of children seriously . . . their competence, determination, dependency or vulnerability does not determine their inclusion or exclusion from participatory processes, but rather informs the way in which their participation takes place" (352). Finally, Graham and Fitzgerald point to a number of ethical implications of those conversations, and emphasize that to recognize child citizens more expansively requires adopting a "standpoint of respect for their views, perspectives and assumptions" (352). What might these forms of recognition mean in terms of the geographies, mobilities, and rhythms of climate change?

A 'PERFECT MORAL STORM'

Accepting the need to move beyond constituting children as human becomings necessitates rethinking how to approach the likelihood that present and pending generations of people under eighteen years of age will inherit and co-produce a climate-changed world—albeit to vastly varying degrees. How, then, to share that knowledge about climate change, and support children to hold themselves and others to account, while protecting their specific rights *as* children? In responding to that question in what follows, it is not my intention to rehearse the numerous arguments about the causes, scope, and characteristics of climate change that exist in a burgeoning literature. Rather, let the starting point be that there is broad consensus that climate change is real, as evidenced by extraordinary and growing amounts of data and information about its ecological, social, and cultural implications. The agreement afforded by that consensus is nevertheless characterized by significant ethical challenges, many of which are usefully examined in a collection of essays introduced by Stephen Gardiner (2010a). In one essay in that collection, Paul Baer (2010) works from the premise that the climate system is a life-support commons and, understood in that way, invokes both the right not to be harmed by any harm that befalls the commons, and the right to compensation for any harm actually done. By reference to an analogon—the ethical and legal debate on pollution—Baer describes several conundrums. One of these puzzles is the spatial and temporal separation of elements in a causal chain. In other words, the *rhythms* that attend climate change involve significant lags and complex patterns of flow and duration—particular gases in the atmosphere, for example. Another and allied dilemma is the uncertainty that typifies the science establishing any such causal chain, especially given that it comprises complex and *mobile* collectivities, some of which harm, and others of which are harmed. Then, Baer writes, and "perhaps most critically, [is] the question of intent and the predictability of subsequent harm at the time that the actions were taken" (25). Developing an acceptable and meaningful index of responsibility will be difficult, he suggests. Present and pending generations of children are implicated in what Baer writes.

In the same volume, Simon Caney (2010) draws on a human rights framework to claim that combating climate change should never violate rights to life, health, and subsistence—the right to prosper and flourish, in effect. Enacting that principle would then require "that the least advantaged—those whose human rights are most vulnerable—should not be required to bear the burden of combating climate change" (172). Children clearly number among such groups and are implicated here, because the *geographies* that underpin their vulnerabilities are both complicated and shifting. Yet enacting an extended understanding of their responsibilities as citizens may, in fact, result in their being so burdened. What might matter most, if children are to be seen as active citizens rather than in-waiting, is the support they are provided to operate in such contexts. This observation folds back into ethical questions raised in chapter one that relate to discipline and conduct, and that implicate practical wisdom to foster flourishing.

Gardiner's (2010b) own essay in the aforementioned collection claims that climate change is a perfect *moral* storm: a confluence of many events, causes, and effects that so aggravates our modes of existence and interaction that it becomes extraordinarily difficult to act. Such constraint—a sense of being *immobilized*—is deemed a moral risk. To this analysis, Gardiner adds an *intergenerational* storm on the basis that climate change is "a seriously lagged phenomenon . . . a resilient phenomenon" and one whose problems are "seriously *backloaded*" (90–1; original emphasis). Backloading, Gardiner asserts, implicates significant epistemic challenges, not least for the political actors and standard institutional arrangements that are predicated on working to short-term goals and providing limited incentives to act otherwise. Significantly, the problem compounds: each new generation "will face the same incentive structure as soon as it gains the power to decide whether or not to act" (92). Climate change impacts also compound and are passed on to future generations—our children—and exacerbated, a scenario that "violates a fundamental moral principle of 'Do no harm'" (93). Gardiner then invokes a third *theoretical* storm, suggesting that the overarching response to climate change is inept. His point is that a generalized incapacity to think through and act in transformative ways in relation to climate change converges with the two other storms to produce moral risk and corruption. If correct, this inheritance is a poor one—and reminiscent of Aristotle's cautionary guidance about the need to consider one's legacy that was noted in chapter one.

Among the key composite challenges of such climate change 'storms' are their present and predicted impacts on a range of mobilities, namely relocation, migration, displacement, and evacuation (Birk 2012). Each of these forms of mobility is caught up in climate change impacts and effects, and all have significant additional implications for cultural understandings of this complex phenomenon. Such issues concern Neil Adger and several colleagues

(2012) in one review of the literature on the cultural dimensions of climate change. In particular, they emphasize the importance of understanding sense of identity in place, place attachment, and sense of belonging to community. They also view such understandings as crucial for developing and delivering effective public policy responses to climate change. Knowing whether, how, and to what extent individuals and communities have cultural mechanisms to cope with the diverse consequences of climate change becomes critically important. Having some comprehension of the ways in which different cultures and sub-cultures may actively or reactively respond to the implications of sea level rise and coastal inundation would be, for example, most valuable to assist with adaptive policy responses and actions on-ground. Given the possible extent of mass movements of people arisinig from such change (not least at the coast), it will be important to discern the value systems that inform present cultural practices, and that influence mitigation and adaptation. This knowledge may be an important predicate to "promote humanist [or other] values that counter exclusive and conformist values" (Adger et al. 2012, 114).

CLIMATE-INDUCED MOBILITIES—A FUTURE FOR ISLANDS?

The entanglements that constitute the geographies, mobilities, and rhythms of climate change are informed by cultural values, and by legal and ethical frameworks, and these matters concern numerous writers focused upon climate-induced migration. Scott Leckie (2008, 18), for example, takes the position that people forced from their homes "must have a remedy available to them which respects their rights, protects their rights and, if necessary, fulfils their rights as recognised under international human rights law". Moreover, where possible, those so displaced should be able to return to their homes without discrimination.

The matter of climate-induced mobilities is also taken up by Jane McAdam and Ben Saul (2010), who call for new theorizations of legal frameworks on the basis that there is credible risk of massive displacement from climate change effects. McAdam and Saul argue that there are significant gaps in understanding what sorts of protection international law affords, despite nearly three decades since the First Assessment Report of the Intergovernmental Panel on Climate Change nominated displacement and mass migration as key challenges. A key question that arises in relation to such possibilities is whether and in what ways human rights and humanitarian laws "apply to all irrespective of whether one is displaced or at home; and there may (or may not) be a compelling policy interest in avoiding the proliferation and fragmentation of legal regimes developed for increasingly specialised sub-groups" (1). Alternatively, there may be cause to extend, adopt, or particularize existing legal principles, or to develop new norms in response to the limitations that characterize present regimes (Stratford et al. 2011). This last option is deemed extremely challenging because there is limited political will and inadequate concern about raising, among affected groups, false hopes or unreasonable expectations that may produce greater levels of vulnerability. The points that McAdam and Saul (2010, 2) make in this regard are punctuated by the observation that people who are "unable to move away from the negative effects of climate change . . . may well be in need of more assistance than those who are more mobile and better able to establish homes and livelihoods elsewhere". At the same time, those who are displaced are "plainly entitled to enjoy the full range of civil, political, economic, social, and cultural rights set out in international and regional human rights treaties" (2) and various forms of law—international, humanitarian, and environmental among them. Notwithstanding this principle, McAdam and Saul point out that there is no present international legal recognition of those displaced by climate change and who may be seen to have express rights or be in need of special protection.

Clearly, children must number among those to whom McAdam and Saul (2010) refer, and those living on islands may be in need of specific forms of protection. A widely sanctioned conclusion drawn from the data on climate change is that "warming of the climate system is unequivocal".[1] In this respect, island specialists working on the IPCC Fourth Assessment Report open their commentary on climate change impacts on islands by noting that their geographies render them highly vulnerable, a status often exacerbated by low adaptive capacity and high adaptation costs relative to gross domestic product (Mimura et al. 2007). Topologically, islands are defined primarily by their coastlines, and many have particular morphological characteristics such as low-erosive rocky shores, high altitudes, or large land areas, that may provide some protection from sea level rise and coastal inundation. However, there is a general consensus that low-lying, high-erosive areas of small islands will be profoundly affected by sea level rise and its attendant effects on water, land, ecosystem processes and services, population distribution, terrestrial and maritime food production, and infrastructure (Figure 3.2).

But no island is likely to be immune from such impacts. Islands with land at higher altitude are also predicted to experience significant change across gradations of ecotone that will affect their productive and natural landscapes. Widespread effects on economic, social, cultural, and political institutions on islands are expected.

A 'worst-case scenario' for islands is that climate change impacts will render entire states uninhabitable, primarily in the Pacific and Indian Oceans. Maxine Burkett (2011, 346) suggests that this possibility will be attended by consequences for international law that are both "great and unprecedented" and that require novel ways of thinking about sovereignty to embrace "a new category of international actors: the Nations Ex-Situ". Certainly, John Connell (2013, 190) grants that sea level rise is the "most anticipated, feared and controversial impact of climate change and global warming" on islands, and its prospect remains deeply politicized. Nevertheless, Connell

Figure 3.2 Funafuti, Tuvalu, Source: Stratford 2005.

is not alone in insisting that caution be used in how we constitute the discourses of island futures. Jon Barnett and John Campbell (2010) describe Pacific Island countries as central players in populist understandings of such changes and effects (see Pacific Islands Forum Secretariat 2012). Properly, I think, they are critical of discourses of extreme vulnerability that tend to constitute islands as defenseless, island states as 'titanic', and island peoples as automatically and necessarily fated to become 'environmental' or 'climate' refugees—a characterization the authors recognize as highly problematic, given no such status exists at international law. As Ilona Millar (2008) notes, refugee status is not afforded to those displaced by environmental causes and appears to trigger no clauses in the United Nations Convention Relating to the Status of Refugees (United Nations High Commission for Refugees 1951) or supplements to it. Moreover, there is silence on climate-induced mobilities in the Convention, and in the United Nations Framework Convention on Climate Change (United Nations 1992; United Nations High Commission for Refugees 1951).

Barnett and Campbell (2010) make an additional point that islanders are often confronted with problems of development and environment, but not necessarily because of islandness *per se*, and they conclude that it is deeply counter-productive to assign to islands the status of incurable vulnerability.

Their grounds for this conclusion are compelling. First, they acknowledge that "there is nothing disingenuous about the SIDS [small island developing states] working with the language of vulnerability . . . [but it is] a powerful concept that cannot always be controlled, and its use can lead to unintended consequences" (166). Second, they delineate one such consequence—a simplistic tendency for others, elsewhere, to blame the 'victim' for living in places over-exposed to varied risks and uncertainties. Third, they note the short conceptual distance between constituting victimhood in such a manner and then invoking externally-sourced heroic interventions in the guise of environmental management, risk reduction for natural hazards, and development projects and processes. Those interventions may be laudable, but remain insufficient in both explanatory and material terms. Furthermore, they elide the North's liability for climate change. In the process, mitigation and adaptation costs, that is "the costs of adjustments to the externalities of the North's pollution [come to be] borne first and foremost by countries like those in the South Pacific" (167). Perhaps most insidiously, arising from these displacements is an expectation that islanders adjust their conduct without expecting the same sort of adjustments from those of the (continental) North. At the same time, we are witness to the problematic constitution of the island into "an absolute, discrete and enclosed space in which climate change impacts and their solutions seem more tangible . . . [a space that comes to] embody not a uniquely located tragedy but the destiny of the entire planet" (Farbotko 2010b, 58).

Inevitably, narratives about this double destiny of islands and the planet[2] are entangled in the production of knowledge and understandings of children's roles in climate change. There is evidence to suggest that in the foreseeable future islanders—and island children specifically—will experience significant privations and be burdened by climate change. Certainly, Sheridan Bartlett (2008) notes that climate change impacts are likely to fall disproportionately upon young people in poverty, not least in coastal places. Those impacts include elevated levels of heat stress and mortality from extreme weather events; water-borne, sanitation-derived, and respiratory illnesses; malnutrition; an increase in vector-borne and infectious diseases; and several corollaries of deprivation and distress, including among adult caregivers. Notwithstanding such challenges, Bartlett insists that children must not be constituted as victims. Their capacity to cope is positively correlated with adults' capacity to engage them meaningfully in active problem solving. In this respect, her conclusions accord with those about resilience drawn by Connell (2013), Barnett and Campbell (2010), and Farbotko (2010b).

The engagement that Bartlett (2008) envisages necessitates widespread institutional and governance reforms. Yet, in discussing a review of fourteen island countries' climate policies, Donovan Burton, Johanna Mustelin, and Peter Urich (2011) demonstrate that reference to young islanders is almost completely absent from national adaptation plans of action derived from the United Nations Framework Convention on Climate Change. In the

Pacific, for example, documents from only Kiribati and the Solomon Islands "clearly indicate a pathway for including children's views in their respective adaptation processes" (9), and only Kiribati stipulates that children are to be involved in relocation discussions. Emergency preparedness plans also fail to mention child separation, child protection, or the provision of temporary classrooms in the event of emergency events—and each of these should be stipulated as determined by the United Nations Children's Fund (UNICEF) in a series of Core Commitments for Children in Humanitarian Action that originate from the 1990s (UNICEF 2010). In response, Burton, Mustelin, and Urich (2011) urge policy-makers to elicit and include children's views; consider culture and community as well as technology when formulating policy; teach in local languages; enable learning by doing—for example, by encouraging weather monitoring and food production; gather and share information relevant to children's well-being and generational concerns; include 'summary for children' sections in reports; and strive to better understand the implications of migration. Thinking about the specificities of *climate change* migration, one is reminded of Adey's (2010b, 13) assertion that, at its core, mobility is a spatial displacement, and that "whether material, electronic or potential . . . Geography forms a window from which to look out and see". At the same time, Adey notes, *motility*—the capacity to be mobile—invokes a range of ideas about citizenship, displacement, mobile politics, and liberty, and includes the appropriation of those things needed to move. At least some of these ideas pertain to children as refugees and asylum seekers (Archambault 2012; Hopkins and Hill 2008).

Some such ideas are also explored in an edited collection on the global-scale mobility challenges posed by climate change. Introducing the essays that comprise the collection, Kirsten Hastrup and Karen Fog Olwig (2012) outline how climate change migration has been constituted numerically and statistically, and been augmented by consideration of the political, social, economic, and moral entanglements of the challenge. Elaborating upon these concerns, Hastrup and Olwig propose that climate change and migration require new ways of understanding *the social imaginary* in ways that account for "the present liquid times [in which] the boundaries are dissolving" under the effects of global economic (dis)order (5).

Yet, arguably, the dilemmas of climate-induced mobility may etch national borders more deeply, and result in significant numbers of people *being denied motility*—that capacity to move. Such possibility invokes both heterotopia and heterochrony: other geographies; different temporal and spatial scales of fixity; and differential rhythms—varied durations of dwelling or displacement or movement; numerous intensities and cycles of change; engagement in shorter and longer term risks, acute disasters, chronic hazards, or single and multiple events; or mutable intervals between incidents. In this vein, consider Thomas Birk's (2012) commentary on the multiple complexions of mobility that have characterized island life. Birk notes, for example, the historical tendency for Pacific islanders to use relocation as a strategy to reduce risk in response to environmental, economic, and social change, and only some of that is likely to be as a result of climate change and sea level rise. Other factors for migration have included local constraints in terms of "food production and the diversification of livelihoods, and through inter-island trade, alliance brokering, and intermarriage" (88). Birk also mentions the importance of work-based migration and remittances, which have become more important over time (see also Bertram 2006). Certainly, most island communities have widely distributed diasporas that work on international, national/inter-provincial, and intra-provincial scales. Many have also experienced 'relocation by persuasion' on the grounds that it will address challenges such as landlessness, joblessness, homelessness, marginalization, food insecurity, increased morbidity, loss of access to common property and services, and social disarticulation.

A MAP OF A DREAM OF THE FUTURE

According to Michael Hulme (2009, xxxvii), climate change needs to be approached from new vantage points; despair is immobilizing, and there is "creative psychological, spiritual and ethical work that climate change can do and is doing for us". Indeed, Hulme proposes that "climate change is an imaginative resource around which our collective and personal identities and projects can, and should, take shape" (xxviii). Again, such observations echo ideas about prosperity, and the realization of processes and systems to flourish, which are of central concern to me here.

In such light, the balance of this chapter focuses on imaginative and creative prospects for a climate-changed future in which young islanders experience resilience. The focus is upon a collaborative project spanning 2008–2010 that was based in the island state of Tasmania, Australia. *Fresh! A Map of a Dream of the Future* engaged young people aged nine to twelve, school educators, academics, community cultural development professionals, and artists, including writers and filmmakers. What follows is a brief analysis of that project, which is indebted to two principles in the United Nations Convention on the Rights of the Child. The first, Article 13(1) noted above, provides for minors to have voice in public and civic affairs (United Nations. Office of the High Commissioner for Human Rights 1989). The second, Article 22, allows that any child seeking refugee status or "considered a refugee in accordance with applicable international or domestic law and procedures shall, whether unaccompanied or accompanied by his or her parents or by any other person, receive appropriate protection and humanitarian assistance in the enjoyment of applicable rights set forth in the present Convention and in other international human rights or humanitarian instruments to which the said States are Parties". The relevance of these Articles will become more readily apparent shortly.

Exploring the range of struggles that typifies participatory research with children and that is partly motivated by such ends, Christine Pascal and Tony Bertram (2009) suggest that active listening to young children is critically important in extending to them the range of rights envisaged in the UN Convention. In this context, empathizing by listening "includes using all the senses and emotions and accessing children's range of communication [which] is clearly not limited to the spoken word" (255). Pascal and Bertram make a point of wider salience: enacting democratic and human rights frameworks in scholarship has long-term implications for children's participation in civic life; such participation should be conceived in multiple ways. The worth of such possibilities informs the values embedded in *A Map of a Dream of the Future*. The chief functions of AMDF, as it came to be known, were to provide opportunities for young islanders to work with people from the creative industries, and consider the ways in which island life under climatically changed conditions might engender opportunities for resilience and innovation. The project involved, in 2008, a public forum on islands and climate change. It was followed in 2009 by a series of education workshops for upper primary and lower secondary children, which required the team to develop a novel climate change education kit for teachers and after-school program leaders to work from. During workshops, children were invited to answer a series of questions about water and food, shelter, transport, and migration (the last of these being a chief focus of the discussion below). They were also asked to consider themselves politically thoughtful and active citizens *now*. In general terms, responses to the questions elicited how children would *prefer* to address climate change challenges. The survey was modeled on the Political Compass, an online tool that plots responses to political questions in quadrants according to whether they are left or right of center, and more or less libertarian or authoritarian. The AMDF survey sought to establish whether children's views were more or less ecocentric or technocentric, more or less libertarian or authoritarian, and generally optimistic or pessimistic. Then, the following year, findings were translated into an installation (Figure 3.3) at the center of the Australian National Regional Arts Festival, *Junction 2010*, in Launceston, Tasmania. Several hundred live plants were clustered and suspended over shallow water in which tread-pads were placed so viewers could walk into the setting, each cluster representing a plot-point in three-dimensional space of the views of nearly one hundred child respondents on climate change as they were graphed in a corresponding virtual world in Google. Thus the circled grid point in Figure 3.4 represents a child whose views on climate change are broadly libertarian-technocentric, accounting for migration, shelter, transport, and food and water. Indeed, most students were more libertarian than authoritarian (upper half), and invested in technocentric solutions to climate change challenges (left half).

It would be intriguing to speculate about whether, how, and to what extent these results reflect Tasmania's relative affluence, constitution as part

Figure 3.3   A map of a dream of the future, Installation, *Junction 2010*, Launceston. Photograph courtesy of Nicholas Low. Reproduced with permission.

of a neoliberal democratic polity, and highly developed and modernized status. Notwithstanding, the team wished to engage with children on climate change in ways that might be enlightening for everyone concerned, and in ways that did highlight Tasmanian children's privileged position. We wished to work in ways that might be empowering for children, especially given how much material on climate change emphasizes vulnerability. We also wanted to provide insights about how participatory arts practice and scholarship might support extended and hybrid forms of citizenship for children.

AMDF was initiated by adults rather than children; the reverse is extremely rare in formal engagements across generations (Malone and Hartung 2010). Nevertheless, the project involved those aged between nine and twelve years of age *opting in* to the work program in class settings and after school programs in regional art galleries. The one hundred or so children live on one of the world's larger island groups, Tasmania, comprising some three hundred and thirty-four members, whose mainland area is approximately 26,216 square miles (62,401 square kilometers), with around 3,355 miles (5,400 kilometers) of coastline—of which some eighty-four percent will be subject to inundation as a result of sea level rise induced by climate change (Sharples 2006). The children belong to an advanced economy; Australia being ranked in the top twenty with a gross domestic product of US$1.23 trillion—roughly $66,000 per capita (International Monetary Fund 2011). Although Tasmanian children may not face the extent or degree

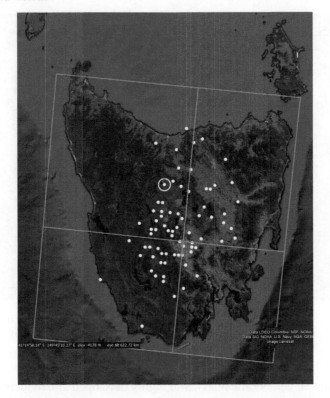

Figure 3.4 Island children's views on climate change. Photograph courtesy of Nicholas Low. Reproduced with permission.

of vulnerability to climate change impacts that children elsewhere will meet, the implications of change are global; mitigation and adaptation require responses that are mindful of that scale of change and its local manifestations. Due thought will need to be given to the predicaments others will face in distant places that may require them to leave their homes and communities, and that will pose international challenges pertaining to sovereignty, territory, and nationhood, which will affect children everywhere. Alongside their counterparts elsewhere, Tasmanians will need to think about how to respond to the relocation needs of those who are displaced.

With these points in mind, in 2009, the AMDF team prepared a forty-four-page book at the center of which is a short story set in the year 2090 about three island children—Ruby, Kené, and 'you'. In workshops, teachers and after-school program leaders were encouraged to read the whole story over several sessions, assist children to discuss the issues raised, and then undertake writing, drawing, and other work in class, prior to completing survey questions. In several parts of the story, which is variously focused on transport, food and water, migration, and shelter, we learn more about the children. An extended extract drawn from the introduction and the section on migration provides a sense of the narrative, and sets a context for the discussion that follows.

May 2090. It's an early morning in May and you, Kené and Ruby are travelling along the beach on your way to school. It's low tide on a beautiful warm day. Waves lap at the shore. Small boats pass back and forth to the island in the bay, powered by sails, paddles and the gentle hum of salt-exchange engines . . .

You're following the coastline, bending and curving past where the old town centre used to be before the sea levels rose. The buildings that once stood there have been pulled down and the materials reused in the new houses that dot the hills. You pass the Tide Monument on your right, floating in the sea to mark where people used to live and work in the old days . . .

All of a sudden a mass warning message beeps in the news system. You check the display. "Whoa," you say to the others. "That's not good. There's a big storm due in half an hour." At the same moment, a second message comes in and the town bells begin to ring. The sound carries across the marketplace . . . The bells usually only ring in celebration but today isn't a special holiday. There can only be one other reason for the bells ringing.[3]

"The refugees," says the man [who is collecting old plastic bottles]. "This must be the people who sent word that their islands have been flooded" . . . You look out to sea and against the storm building on the horizon you can make out some . . . boats. They've been battered by the storm and are flying strange white and green flags . . .

Kené tells you, "That's the Cli-Ref flag—the flag climate refugees fly . . . If you're flying it other people will come and help . . . Quick, let's get down to the beach to meet them before the storm hits!"

. . . The rest of your class stands on the pier along with most of the townspeople. Everyone is curious to see the new arrivals. It's been a long time since climate change refugees have arrived in your town and your friends have mixed feelings.

"I wonder where they will all be living? Our house is full up!" says Ruby.

"But there's always ways to make more room," you say . . .

"I bet they've got ideas for building houses that we've never even heard of," says Kené.

As another boat slides into the dock a boy on its deck sees you and your friends watching. He looks scared. Ruby waves at him and shouts "Hello!"

He smiles and waves back, looking a little relieved . . . You step forward and take the antique bottle [you found earlier on the beach] out of your pocket. You hold it out to the boy and place it in his hands as a welcoming gift. His smile turns into a grin as the first drops of rain begin to fall . . . (Low et al. 2010, various pages)

The migration story conveyed above was one section of the larger story that children were invited to consider before sharing what they might do in response, and prior to exploring why and under what circumstances people leave their places of dwelling to go elsewhere. Teachers and after-school program leaders could invite children to create welcome posters highlighting the benefits to incomers and recipient communities of in-migration, and then consider arguments for and against accepting refugees into communities. Older children could run a debate for and against a migrant resettlement program in their community. All children were asked to write poems about being climate refugees in which they described the experience of leaving home and moving with other members of their community to a new place.

Teachers and after-school program leaders could then invite children to complete survey questions developed by Nicholas Low, the project's lead artist, and an educational psychologist named Tim Cotter. In response to the migration story, forty-two boys, forty-one girls, and four whose gender was not specified responded to six issues, and rated six associated and paired ideas about how to respond to those same issues. Adapting parameters of the Political Compass, certain of the ideas listed below were presented as statements to the children; some are authoritarian (Ideas 1, 3, and 6), others broadly libertarian (Ideas 2, 4, and 5); some are technocentric (Ideas 7, 10, and 11), others reflect an ecocentric perspective (Ideas 8, 9, and 12). The children were asked to rate the statements according to whether they thought them 'terrible, bad, good or brilliant', and they tended to favor technocentric and libertarian ideas. One hundred and fourteen additional comments were logged, an uncorrected selection of them below made primarily on the basis that they illustrate the breadth of response or are especially evocative.

The first statement presented for consideration was that, *over time people in some areas of Australia will have to leave their homes as a result of climate change.* The paired ideas were (1) that the government orders people to move away from places suffering the effects of climate change—which is authoritarian, and (2) that if their homes become unsafe, people organize into groups and choose whether to stay or go—which is more libertarian. Fifty-two children thought the first idea bad or terrible; seventy-one rated the second either good or brilliant. Twenty-four offered their own ideas or comments, including the following:

Technology and financial support [should be] given to enable these places to become livable in sustainable ways.

The people get to choose if they want to stay or not.

The government makes houses for people that are close to the beach and can be affected by climate change and the people can choose when they leave.

What if climate change does not happen?

The second statement was that *people from places such as Tuvalu [which the children had learned about in class], that may be lost to rising seas, may need new countries to call home.* Here, the paired ideas were (3) that governments in wealthy countries will be forced by the United Nations to take in climate refugees and (4) that it is up to each individual refugee to try and gain acceptance to a new country. The term refugee, while contentious, was deliberately chosen because of the widespread use it has in Australia, and the controversy that is attached to it. Views were split in relation to (3), with twenty-nine thinking it bad or terrible and fifty-seven deeming it good or brilliant. Similarly, thirty-five children thought idea (4) was bad or terrible, and fifty-two thought it good or brilliant. Among the twenty written responses were the following ideas:

Creating islands out of waste material that can be used to provide a similar life for these people if this is what they choose.

If people just lost their homes they should be allowed where ever they want.

The governments should let refugees in there contry [sic].

People should be honest and trie not to get into the countrys illegaly [sic].

The third statement was in the form of an assertion that *climate refugees and existing communities will need to live together.* Idea (5) posited that climate change migrants start their own communities, build their own houses, and make do where they are relocated, and forty-five children thought that this was good, another eleven deeming it brilliant. Idea (6) was that the government provides climate change migrants with houses in existing communities; fifty-two thought this a good idea, and a further sixteen thought it brilliant. Among seventeen additional comments were the following:

Refugees decide to live in communities or to build their own but either way there are creative swaps of knowledge and culture between these groups.

Climate change migrants should make there [sic] own communities but they have to do what the government says.

At the side of a city, there will be a place for the traditional houses, according to the migrants religion.

Migrants can live with us they are the same as us.

The fourth statement was that *some areas will be at risk of erosion and flooding*. Idea (7) was that people use their money to build expensive, high-tech sea barriers to protect their land, and idea (8) was that they pack up their belongings and move. Just over half of the children (47) thought (7) sound, and (8) not so (48 children). Twenty-two offered comments, among which were these:

> All land needs to be protected but financial support is given to all to do so.

> If the sea rises too high, the council pays money to move your house inland.

> If people pack and move away from the sea they will be safer.

> I would hate to move to a different place.

The fifth statement raised the prospect that *people from climate-affected islands will need somewhere new to live*. Idea (9) was that island people will have to live nomadic lives, traveling around the world and living in different places along the way. Idea (10) proposed that artificial islands be created to replicate the ones that are lost. Responses were starker here, with fifty-nine children deeming the first idea bad or terrible, and sixty-five suggesting the second good or brilliant. Of the seventeen who shared their own ideas, four wrote the following:

> I think that some artificial islands should be made but some people should move.

> I don't think they can replace the islands that are lost.

> Make earth one big country.

> The people who live on remote islands will move to the mainland.

The last statement was that *climate migrants and people from their new home will have to learn each other's languages*. Idea (11) was that automatic translation devices let people understand each other. Idea (12) was that people live together to learn each other's language. Both ideas were largely seen as good or brilliant, seventy-six supporting translators; sixty-eight learning other languages. Fourteen additional comments were provided, including these:

> People need to understand each other.

> If other people learn other languages they will get along better.

> People should learn other languages so it's a lot easier to understand each other.

> There will be language schools across the country.

It becomes clear that the children had diverse views on climate change migration; the differential roles, rights, and responsibilities of governments, individuals, and communities; the distinctions among displacement, relocation, migration, or evacuation; the utility of resilience and inventiveness; and the varied ways of being in place, on the move, and in dialogical relation. Those views also reflect different ethical and political positions, naïve and sophisticated approaches, levels of inventiveness and reaction, and optimistic and pessimistic mindsets. They exhibit, in short, the sorts of positions one might expect from a wide range of citizens exercising their voices. Two of the comments offered by children participating in AMDF are, for me, especially illuminating: that governments should let refugees into their countries, and that asylum seekers should be honest and not enter countries illegally. These invoke again the challenges that attend the observance of principles enshrined in the UN Convention on the Rights of the Child and pressing challenges about the relationship between 'honest' approaches and those compelled by a sense of looming peril and desperation. Such comments on the part of two children seem to cut to the heart of the matter—at least in Australia, where the 'management' of asylum seekers (otherwise labeled illegal immigrants) has become deeply politicized and highly divisive; at time of writing there are over six thousand such people in detention here.

DREAM ON?

I began this chapter by reference to the work of William McTaggart, which I deployed as emblematic of the multiple geographies, mobilities, and rhythms of climate change. At this point, it seems useful to reconsider the significance of the group of children in *The Storm*: on the basis of the discussion elaborated in this chapter, any suggestion that their quiet vigil is passive warrants redress. Imagine, instead, how actively engaged they may be: feeling, thinking, forming views, and absorbing the scale and rhythm of natural and social entanglements; reading the geographies of their place; learning to understand the adventures and misadventures that typify the life-course; and coming to appreciate how deeply embedded we are in what Tim Ingold (2005; 2010) has termed the weather-world.

Climate change, too, invokes a range of storms: moral, intergenerational, and theoretical. It is a challenge of immense complexity, especially for children and in relation to increasingly unhelpful conceptions of them as human becomings or as political actors in waiting. Alternative views are likely to be critically important in unsettling ideas of children's subordination, and reframing the ways in which citizenship is constituted with and for them, including those that register the importance of children's geographies, and that *recognize* them.

Climate change is a challenge of undeniable significance for islands and islanders, especially in relation to resilience and the fourfold consideration of displacement, evacuation, relocation, and migration. Noting again the extent of evidence that shows credible risk of massive displacement from climate change, jurisprudential and other scholars are considering how to address questions pertaining to territory and sovereignty in terms of the Nation Ex-Situ and other legal innovations or inventions. Alongside these various considerations is the very real possibility that significant numbers of island (and other) peoples, children among them, will experience the geographies and rhythms of climate change most pronouncedly in terms of mobility, and that—equally testing—many will not have the capacity to move, and some may choose not to (Stratford, Farbotko, and Lazrus 2013). Such circumstances call for new social imaginaries, new vantage points, and new forms of engagement with children. In my view, collaborative and participatory arts programs provide at least one mechanism by which to co-produce novel and empowering cognitive, affective, political, and real maps for the future. They enable us to continue to consider—as we move through the life-course—how to conduct ourselves, how to govern; and they invite us to consider how we constitute the conditions by which to flourish.

NOTES

1. Global "average sea level has risen since 1961 at an average rate of 1.8 [1.3 to 2.3] mm/yr and since 1993 at 3.1 [2.4 to 3.8] mm/yr" (Pachauri and Reisinger 2007, n.p.). Taking the average as base, that is nearly fifty-eight millimeters (over two inches) from 1961 to 1992 and sixty-five millimeters (over two and a half inches) from 1993 to the present time.
2. Here one recalls how, when Earth was first seen from space, its islandness in a sea of black space inscribed into a generation's psyche twin ideas about the planet's singular and special status and its limits (see Cosgrove 1994).
3. Tolling bells invoke all manner of visceral reaction, their rhythmic tone often a powerful harbinger. One can imagine the villagers in McTaggart's painting of *The Storm* being roused from their beds by the clangor of ship or school bell, just as the three children here are also rallied to action. The visceral politics of sound is powerfully drawn out by reference to another climate change exemplar—protest parades in a New South Wales coal region—by Gordon Waitt, Ella Ryan, and Carol Farbotko (2014, 298), who remind us that "sounds play a crucial—but often under-recognised—role in mobilising bodies as individuals or collectively".

4   Grind and Trace

Around 2006, in Manchester, Loughborough, and London, artist Lottie Child began using ephemeral experiences of play she calls 'street training', the purpose of which is to subvert how people engage with urban spaces. Child (2010, 86) describes her work as akin to a martial art requiring "regular attention and practice . . . [to] improve joyful and safe behaviour, thus enriching environments and our experiences of them". The idea that spatial relations and sense of place involve delight is a constant in Child's work, which invites adults to reengage with play and learn from the joy exhibited by young people. Street trainers thus seek to understand spaces and places afresh, becoming alert to how they are monitored, privatized, and commercialized. Accordingly, Child (2007, n.p.) stresses that "spontaneous small-scale happenings in public space" are a "distributed assault on the pervasive culture of fear and cynicism that we otherwise perpetuate and normalise as we constitute our urban public spaces". Such motivations have prompted street training with strangers, students, police, risk analysts, martial arts practitioners, urban developers, architects, and artists. Integral to the practice are playful movement and exploration, creativity and expression, and courage and athleticism. Participants draw on pavement with chalk, dance, guerrilla garden, walk blindfolded, climb, slide down handrails, or lie on railway station floors doing 'nothing'—this last act highlighting the utility of the interval and interstice, those gaps of indeterminate length which create space for adventure (Figure 4.1).

In one session, Child (2009) and several other women wander across an empty rooftop car park in Manchester, playing in puddles, and a clip of the event on YouTube captures Child's intent (Figure 4.2, taken at 1 min 41 sec). The initial focus is on details at the interface of limb and puddle. Pivoting from an unseen hip, a sneakered foot descends to water that adheres to crenulations of the sole and then is repeatedly drawn out of the reservoir by twists and turns which create fronds rupturing the border of the puddle. Across the camera's microphone, the slurping mixture of shoe, water, and ground intersperses with wind-rustle and metallic clangs—rhythmic beats of a rope on a flag pole. Sky and cloud reflect in the water. Nearby, drainage holes squat in the concrete, and white lines mark spaces where cars might

Figure 4.1 Street training I. Photograph courtesy of Lottie Child, 2004. Reproduced with permission.

Figure 4.2 Street Training II. Source: Lottie Child, 2009. Reproduced with her permission and sourced from YouTube (Manchester Street Training 10).

otherwise curtail play. A final flourishing side kick from the foot precedes a pan-shot out to the other participants. One taps cowgirl boots in the middle of another puddle. A second traces the border of an adjacent patch of water, walking toe-to-heel. In the largest of the puddles, a third strides from one dry 'island' of slightly-elevated bitumen to another. In even tempo, the fourth walks backwards from another patch, leaving footprints on concrete. The last speeds to the largest puddle, leaps and splashes, laughing as she does. Individually and collectively, these women settle into the childlike by means of once-familiar embodied practices that might epitomize a child's play on a rainy day. In doing so, they produce varied rhythms: patterned and recurrent movements, and metered routines across a landscape whose usual entanglement in complex transport geographies is interrupted.

What possible consequence could such activities have? Certainly, none of the practices in which the women engage even vaguely resembles the rationality of normal(ized) journeys of *homo mobilis* or *homo economicus* (Van Geenhuizen and Nijkamp 2003). Yet as Tim Edensor (2010c, 71) implies in relation to walking (one regulated form of mobility that Child seeks to unsettle), an alertness to the unexpected refreshes the mundane and habitual, attunes us to and mobilizes place, and engenders a sense of belonging and routine—all of which bolsters what he celebrates as "the incipient tendency to wander off score".

Other and allied wanderings are the focus of this chapter. Prompted by playfully subversive engagements of young people in public space, I want to explore three questions. *How might existing scholarship on walking, for example, inform a reading of skating and parkour and their geographies, mobilities, and rhythms? How do older adolescent and young adult skaters and traceurs use urban spaces and produce ludic geographies contoured by particular movements and rhythms? How and with what effect on flourishing are their practices entangled in specific modes of governing and understandings of settlements that appear to render them unwelcome?* Addressing these questions invites consideration of the meaning of 'young' among those labeled adolescents, teenagers, and young adults. In Australia, from whence I write, that span of years corresponds to one used to determine eligibility for the federal government's Youth Allowance (Australian Government. Department of Human Services 2013). Elsewhere, child psychologists at London's Tavistock Clinic are guided to view adolescence in three stages from twelve to fourteen, fifteen to seventeen, and eighteen to twenty-five (Wallis 2013). Like childhood, adolescence and young adulthood are constituted as overlapping categories—biological, social, cultural, legal, political, and economic (see, for example, James 1986; Panelli, Nairn, and McCormack 2002, footnote 2). At numerous scales, understandings of adolescence and young adulthood are delineated by international conventions, national and subnational laws, or municipal regulations. Recall from chapter 3 that under the United Nations Convention on the Rights of the

Child, childhood extends until eighteen unless national governments deem otherwise. Either way, one of the unifying understandings of adolescence is that it marks a period of growing responsibilization and apparent liberty. It is often characterized by concern and suspicion among adults, and naturalized as problematic. In Sandy Nicholson's *0 to 100 Project*, for example, twelve-year-old Lochlan Finney describes his age as full of "Responsibilities. Like, you have to do all this stuff. You know, work around the house. Got so much homework when you just start high school. It's horrible". At fourteen, Campbell Emerson writes "I have to be responsible for way more". Jacob Fischer-Schmick, fifteen, observes that "When you're a teenager, there's a lot of kind of prejudices against you about your age".

These boys may come to internalize, reframe, co-produce, or reject what is expected of them as (adult) citizens. In this vein, Mitchell Dean (1999, 166–8) suggests that the constitution of *active* citizenship means considering the risk "of physical and mental ill-health, of sexually acquired disease, of dependency . . . of being a victim of crime, of a lack of adequate resources in retirement, of their own and their children's education, of low self-esteem and so on". Dean maintains that calculations of risk involve the "dangerousness of certain activities . . . places . . . and populations . . . [and those which] traverse each and every member of the population and which it is their individual and collective duty to control". The effect of such calculations is the emergence of new technologies of agency, particular understandings of prudential citizenship, and the targeting of particular populations. Of note among Dean's observations are those about the rise and spatialization of ubiquitous practices of government constituting obligatory passage points to get citizens "to agree to a range of normalizing, therapeutic and training measures". Little wonder, then, that Lochlan, Campbell, and Jacob focus on youthfulness in terms of responsibilities and prejudices. But where the three boys concentrate on elements of growing up that somehow seem to disregard play, two pastimes they may engage in as young men—skating and parkour—require of practitioners specific modes of conduct that oblige a sense of responsibility *and* playfulness. The dilemma is that these *disciplines* are not widely seen as such; indeed, they are often represented as 'nonsense' or dangerously disruptive, and are regularly marginalized from cities and settlements. Such responses are part of a larger quandary about the geographies of youth.

YOUTH CULTURE AS "CONSTELLATIONS OF TEMPORARY COHERENCE"

Foundational research on the geographies of adolescents includes that by Gill Valentine, Tracey Skelton, and Deborah Chambers (1998). Introducing *Cool Places: Geographies of Youth Cultures*, the authors spatialize established understandings about the contingency of the formative years. They note the capacity for children to grow or shrink developmentally rather than chronologically: thus on any given day, and under varying circumstances, Lochlan, Campbell, or Jacob might have moments where they conduct themselves in ways discernible as six or twenty-six.

Susie Weller (2006) argues for geographical research specifically about and for teenagers, and does so on the basis that there is need for a more nuanced analysis of teens' liminal status and the manner in which they relate with other generations. Faith Tucker (2003), too, considers how generational dynamics are played out for twenty-seven ten- to fifteen-year-old girls in rural south Northamptonshire. Tucker worked with the girls to understand how they experienced and made place for themselves. She found young people are often highly visible in rural settings, and are often told to leave or 'move on' from public spaces in villages and towns. Whereas some of the girls she worked with comply and retreat, others "actively contest the use of space" (117), operating according to tacit, internalized flows of movement-occupation-movement, navigating their landscapes to account for their own and others' mobilities and rhythms. Tucker's observations reveal the differential movements and territories of dominant and rival groups, and expose various tactics to deal with those such as avoiding others' gaze or turf, or walking fast through or detouring around certain areas. The girls showed how it is possible to use the presence of parents with younger children to occupy spaces they would not otherwise have access to, the company of adults changing the dynamics between groups of teens. Tucker's work usefully suggests that both *intra*generational *differences* and *inter*generational *similarities* exist. The unsettling effect of her work also invites us to ask how public spaces might be better designed for multi-generational occupation on the premise that such spaces will enable new geographies, mobilities, and rhythms that support well-being and sense of place. In parts of this chapter, I hope to touch on such matters.

Useful background to my own work here are Craig Jeffrey's reviews on the geographies of young people. In the first of three reports for *Progress in Human Geography*, Jeffrey (2010) examines a number of structural issues affecting education and employment, outlining different ways of conceptualizing youth prior to the nineteenth century; the influence on understandings of age and generations of both liberalism and romanticism; and the invention of adolescence and then the teens. Jeffrey then considers how, since the 1980s, there has been a growing emphasis on the pathways, forms of navigation, and trajectories that young lives take. At least some of the research he mentions disrupts prevailing ideas about an uncomplicated progression from youth to adulthood on the basis that, in some settings and cultures, the size of populations relative to opportunities means that some will be "unable to acquire the social status of adulthood" (497). Such insights problematize what Jeffrey refers to as "normative, teleological assumptions of life stage models" (498). In supplanting the transition model of aging, for example, Jeffrey draws on Jennifer Johnson-Hanks' (2002) ideas about vital conjunctures to which I referred in chapter one. He emphasizes the idea that "structures contingently combine to shape actions in particular spans of time, where structures are imagined as mutually sustaining schemas and resources that enable and limit social action" (498). Jeffrey also suggests that vital conjunctures aid consideration of the geographies of children and young people on three grounds. First, the term draws attention to the uneven ways in which challenges manifest for children and young people both socially and spatially. Second, it invites close examination of the social and spatial structures influencing particular situations. Third, it provides a mechanism for thinking across and politicizing spatial and temporal boundaries in terms of *critical durations*—terminologies that underpin Lefebvre's rhythmanalysis.

Jeffrey (2012, 245) later extends his review of the geographies of children and youth, focusing on how young people's diverse practices illuminate "the spatial nature of resistance, multiple forms of agency, and the social and often mischievous character of subaltern action". He notes certain effects of binary constructions of young people as "heroes and zeroes" (247), most obviously in the media, and asserts the need to better understand and value how "children and youth navigate plural, intersecting structures of power" (246) from the micro-politics of the home to the global politics of international social movements. Importantly, these navigations are strategic as well as tactical—oriented to the long-term and not just to the challenges of survival; in this sense they *feel* almost Aristotelian. They also implicate diverse spatial practices and the methods used to claim or extract oneself from place (Pain 2003). Here, Jeffrey acknowledges that young people are embedded in "reactionary strategies that sustain and replenish established power structures" (249) as well as radical projects, none of which assumes that they are unthinking or manipulated. Either way, and returning to Johnson-Hanks' (2002) discussion of vital conjunctures, the idea is that young people are highly active in varied forms of mobilization and in producing and reimagining place. As Doreen Massey (1998, 124–5) suggests, youth culture may be more about networks than "the neat packaging of space [or time] into a hierarchy of scales" and might be more productively conceived of as "*constellations of temporary coherence*" (original emphasis). This point serves as the launchpad for Jeffrey's (2013) third and final report calling for more research to understand key moments of change in young lives, and to better comprehend territorialization, resistance, securitization, and other spatial dynamics. He valorizes the significant long-term investments young people make in diverse forms of political agency. Jeffrey then deploys an ecological metaphor to explain protest in terms of technologies, materials, places and complexity, extemporaneity, and changeability; these ecologies of protest can "provide a framework for examining the rhythms, regulations and improvisations that constitute politics in practice" (150).

Notwithstanding the differential meanings that properly are attributed to age, and recognizing the confounding effects of profoundly aging activities such as child labor, at some point the vast majority of young people are expected to make a transition from 'childish pursuits'. In the *0 to 100 Project*, for example, Josh Ferreira, aged eighteen, remarked of his age, "Yeah, that's when life pretty much begins". Arianna Claridge-Change, aged twenty-one, more lightly commented, "Nothing's too serious yet. I don't want it to be ever". Sophie Olsson, at twenty-three, observed "I need to grow up quite soon. Which is a bit scary". Bethan Evans (2008) suggests that in such transitions, young people experience particular kinds of geographies implicating three modes: being nothing, having nothing to do, and having nowhere to go. Implicit in these notions is the constitution of youth as a *movement* to a stable state (adulthood) typified by being someone, having things (of importance) to do, and having a place (physical, ontological). Evans's counterproposal is for "a consideration of the way that young people's lives are connected and bound to others across a range of scales" (1676). The effect is to highlight social and spatial interdependencies in ways that mark the politicized contours of different ways of being in the world. It will become apparent in what follows that skating and parkour are two forms of movement in practice in which are vested significant aspirations in that respect.[1] Arguably, those aspirations often gain expression in terms of play.

ON BEING PLAYFUL

Craig Jeffrey's (2013, 151) work emphasizes the need to consider young people's crucial roles in geopolitical life and affirms how a focus on them "might illuminate wider debates on space, place and scale". What might be the relationship between such concerns and others seeking to understand the critically important functions of play? Thought has been given to this question by Tara Woodyer (2012) in a paper addressing academic engagements with the ludic: what it means to be playful through the life-course. Of particular interest are Woodyer's accents on the relationship of play to geography and the manner in which playful spatial relations inform "the cultivation of a mode of ethical generosity" (313). Engaging with Woodyer's work is a final important step leading to a detailed discussion below of the geographies, mobilities, and rhythms of skating and parkour and their capacity to influence modes of governing and opportunities to flourish. Woodyer outlines certain common understandings of play. In the first view, play is understood as a developmental tool serving other and greater ends—not least opportunities for 'human becomings' to rehearse 'real' life; thus childhood is constituted as an interval prior to the 'business end' of life. Usefully, she ponders the incapacity of such a perspective to explain why play remains attractive to adults. The second view positions play as voluntary, superfluous, having its own disposition (the 'not real'), disinterested (outside of any imperative to satisfy needs), secluded and/or localized and/or typified by specific duration, and characterized by internal rules placed "beyond the rational auspices of Western society" (315). To these perspectives, Woodyer adds a third way of thinking about play's ambiguity. Among the works on which she draws is Brian Sutton-Smith's monograph on the same subject, which repays careful reading. Sutton-Smith's (2009) framework includes mind-based or subjective play (daydreaming, for instance), solitary play (such as hobbies), playful behaviors (playing tricks, putting something into play), informal social play (travel, dancing, roller-skating), vicarious audience play (television, spectator sports), performance play (playing music, play-going), celebrations and festivals, contests (athletics, chance, martial arts), and risky or deep play (caving, skateboarding).

Woodyer also refers to Csikszentmihayli's ideas on the importance of *flow* as something that may typify both work and play. In the sense intended, flow is an all-consuming mindfulness, a presence in the moment or, in other words, an interval of unspecified length that, in my reading, is also biopolitical. Woodyer attributes the idea of play's political status to Cindi Katz's (2001, 2004) work on power relations and social roles reproduction, drawing on Stuart Aitken's idea that play can be both radical and resistant. Aitken (2001, 180) writes that the:

> control that is exercised over young people's play through, among other things, supervision and the design of play spaces, constrains meanings as well as practices . . . Thick² play is thinned and then squeezed out. This is so because play, at its most radical and important, is a form of resistance. Resistance is squeezed out . . . Some behaviors that produce today's spaces for young people are reasoned and increasingly market- and outcome-oriented. Play is not. Transitional spaces are about the kind of imagination and potential that an outcome-oriented world stifles.

Woodyer extends these ideas by additional reference to Ben Malbon's thesis that resistance can be conceived as playful vitality. Malbon (2002, 181) examines clubbing and club culture on the basis that insufficient attention is given to the role of play in everyday lives and that play reveals much about "complex notions of identity and identifications", not least in terms of practices such as dancing in which young people's "playful vitality is a form of oppositional practice in which the primary scale of operation" is the body (154). Woodyer (2012, 319) develops these arguments to suggest that if play is vital (in at least two senses of the term) it is also affirming and, so constituted, "can stimulate a generosity of spirit towards others . . . the vitality emerging from it encourages one to be more responsive to others" through patterned repetitions, improvisations, and intense engagement with the idea of 'becoming-with' others, technologies, and places. Such possibilities inform the cultivation of ethical generosity: "by refracting aspects of society, play is a vehicle for becoming conscious of those things and relationships that we would otherwise enact or engage without thinking" (322).

In this regard, the influence of Michel de Certeau seems clear. In particular, his work *The Practice of Everyday Life* is prototypical in politicizing an ethical generosity of the kind I am seeking to elaborate here. The work is dedicated by de Certeau (1984, v) to "the ordinary man [sic]. To a common hero, an ubiquitous character, walking in countless thousands on the streets". Later, he lauds the "ordinary practitioners" he observes who, down below the corporate heavens of Manhattan's finance district, "walk—an elementary form of . . . experience of the city . . . [their] bodies follow the thicks and thins of an urban 'text' they write without being able to read . . . make use of spaces that cannot be seen . . . each body is an element signed by many others" (93). Engaging with their perambulations, de Certeau seeks practices alien to the geometrical geographies of the panoptic grid that invoke other spatialities—migrational and metaphorical—from which he might discern new forms of conduct with new outcomes for how to live in the world.

Others build on such insights. Hayden Lorimer (2011), for example, demonstrates how walking is a distinctive and meaningful cultural activity typified by traditions, routes, regions, festivals, pathways, renewal of bonds with place, or pilgrimage. Walking is made potent "by the physical features and material textures of place" (35); it constitutes "an embodied space where searching questions are considered, and sometimes answered" (38), often in ways that reinforce, step-by-rhythmic-step, a sense of well-being in the world, and that give rise to invention, agency, interrogation, or resistance. Jennie Middleton (2010), too, focuses on the ways in which walking is rhythmic, practiced, enacted, embodied, sensed, mediated, contextual, mundane, heterogeneous, traceable, mappable, and encoded. Her focus is on the relationship between walking and Guy Debord's (1956) idea of the *dérive*, an unplanned journey or mode of drifting usually in an urban setting, in which one is affected by architectures and geographies discovered and discerned *en route* and re/misappropriated—in French, *détournement*.

Drifting is profoundly enmeshed in the geographical, mobile, and rhythmic, Debord (1956, n.p.) describing it thus: "one or more persons during a certain period drop their relations, their work and leisure activities, and all their other usual motives for movement and action, and let themselves be drawn by the attractions of the terrain and the encounters they find there". In ways that anticipate Lefebvre's (2004) concerns with rhythmanalysis, Debord suggests in the same work that the average span of "a *dérive* is one day, considered as the time between two periods of sleep. The starting and ending times have no necessary relation to the solar day, but it should be noted that the last hours of the night are generally unsuitable for *dérive*". Debord politicizes this psychogeographical practice, pointing to the utility of experimental modes of drifting, old maps and aerial photographs in hand, to create new maps of influences in which the aim is "no longer a matter of precisely delineating stable continents, but of changing architecture and urbanism". For Debord (1967, 13), *dérive* and *détournement* are chief means to resist modern conditions of production in which life is best viewed as a *weltanschauung*, a world view "transformed into an objective force".

Edensor (2010c) also demonstrates how walking's rhythms are subversive, elaborating on the ways in which rhythmanalysis can illuminate certain spatial qualities, sensations, habits, and temporal patterns. He suggests that walking produces a mobile sense of place engendering spatial belonging, routine practices, and social relations, each walker becoming "one rhythmic constituent in a seething space pulsing with intersecting trajectories and temporalities" (71). Each is influenced by standards of dressage, strictures pertaining to the allocation of space as private or public, and forms of regulation that privilege browsing and gazing, practices entangled in the co-production and experience of landscapes of consumption. Finally, Edensor notes that adopting disciplined techniques can heighten the experience of walking. He describes the highly trained body as one no longer fixed on conscious self-management and thus as capable of "unreflexive disposition that may allow moments of eurhythmy to emerge, wherein the body is open to external stimuli and thoughts may turn to fantasy and conjecture" (72). A corollary of discipline in walking is the capacity to improvise and, as established above, improvisation folds back into notions of play, and play into vital conjunctures and productive forms of resistance. Such matters are profoundly important to the ways in which skating and parkour are constituted and come to inform the geographies, mobilities, and rhythms of certain young people.

GRIND AND TRACE

In my teens, I tried skateboarding with two girlfriends on a suburban street in the Adelaide Hills in South Australia, and twice landed hard, ending up on reasonably large doses of analgesic for a severely bruised coccyx. Like many girls, I switched to roller skating and then rollerblading in rinks and parks, and on neighborhood sidewalks (Khan 2009). Estimates from many sources suggest that girl skateboarders comprise under ten percent of all skaters. I remain in awe of girls who stay with it (for example, see work about the Quebec-based Skirtboarders 2009), and have watched with interest as increasing numbers of girl and women skaters and surfers critique the discourses and representations that would seek to marginalize or commodify them (for example, as reported by Pfeifer 2013). Much later, in 1997, I began observing skaters in a public square in Hobart, the capital city of Tasmania in Australia, and later studied skating in inner areas of other capital cities around the nation. Then, in 2000, I was commissioned by the Tasmanian Government to undertake a statewide community consultation in the lead-up to the national legalization of 'small-wheeled devices' as forms of transport on certain sidewalks and roads with no center lines, a legislative change given effect in 2002. I became deeply interested in how young men in particular were constituted more often as 'zeroes' than 'heroes', grew concerned about such binary constructions, and was startled by the vitriol directed at them by others—including in interviews with me. Reports and publications from these studies were among the first on skating from Australia (Stratford 1998b, 2000c, 2002; Stratford and Harwood 2001). Their aim was to contribute insights about skating, urban governance, and the place of young people in public spaces. The work done led to the consideration of social capital, sustainable transport, and sustainability and community more generally.

For several years, as I focused more on studies of local and regional communities and island studies, I kept a weather-eye on the literature on skating and on skaters in Hobart, participating in the design and development of two parks, watching the slow and steady appearance in the landscape of metal cleats and nodules, gravel, lawn, and mesh—fixtures often dubbed 'skate-haters'. I noted the ways in which skaters were steered and, in turn, how they resisted those design 'solutions'. As the essays in this book began to take shape, my thoughts returned to skating and to observations of skaters in the inner city. Simultaneously, I started to hear more about parkour, and although I am told there are *traceurs* in Hobart, I have never witnessed any in action. Either way, I also turned my attention to what has become a significant literature on parkour—meant both in terms of the number of publications and the disciplines from which they have emerged. Here, then, my intention is to provide only brief commentary on the histories of skating and parkour, which are well-documented. Thereafter, I draw on two widely available films to consider the geographical, mobile, and rhythmic elements of skating and parkour as distinct disciplines practiced by young people whose interstitial engagements with streetscapes invite generosity from those of us who do not, ourselves, grind and trace.[3] My contention is that from such generosity might emerge more vital and playful spaces and places.

Skating emerged on the West Coast of the United States during the 1950s. According to Chihsin Chiu (2009, 27), when the surf was poor, young people would remove the T-bars from their scooters and use them to surf first streets and then "deserted swimming pools, drainage channels, and school-yards". By the 1970s, with skaters appropriating other and more populated public spaces, municipal governments started to construct skate parks in response to the growing tensions between skaters and non-skating citizens, especially those with commercial interests in business districts. That management response to skating became standard during the 1980s, and again in the 1990s following a period of decline in park provision resulting from the burgeoning costs of insurance and construction. Over time, skate parks have been the subject of varied analyses about young people's ethical legal status (Carr 2010, 2012); the relationship between skate parks and health and injury (Dumas and Laforest 2009); the ways in which skaters and non-skaters interact in parks (Goldenberg and Shooter 2009); and the relationship between skate park provision and neoliberal governance, particularly in terms of the privatization of public space (Howell 2008; Vivoni 2009). Since the early 2000s, additional design efforts have been used to steer skaters to or from particular public spaces or constrain their movement outside parks—either by means of design—such as the aforesaid skate-haters, or by the use of bans and municipal ordinances and fines (Brunner 2011; Woolley, Hazelwood, and Simkins 2011).

Several authors have investigated the ways in which skaters have continued to appropriate the streetscape as a form of resistance and critique (Borden 2001; Khan 2009; Stratford 2002; Vivoni 2009); others exploring skating as part of consumer culture (Dinces 2011; MacKay and Dallaire 2013). Little wonder then (and appropriately so I think) that, like parkour, skating should be read in the terms afforded by Debord (1956) as *dérive*, *détournement*, and spectacle (Sharpe 2012). Indeed, Debord (1967, 12–13) argued that the whole of life is an "accumulation of spectacles . . . the outcome and the goal of the dominant mode of production . . . the very heart of society's real unreality . . . the omnipresent celebration of a choice *already made* in the sphere of production, and the consummate result of that choice". In this sense, street skating is entwined in the calendrical rhythms of the production process for skate clothing and gear; for example, color palettes and designs are determined (that is, already made) several seasons before products are released to retail outlets for skaters to 'choose' from. Debord also contended that the "device of *détournement* restores all their subversive qualities to past critical judgments that have congealed into respectable truths . . . [and, as] the fluid language of anti-ideology . . . [*détournement*] mobilizes an action capable of disturbing or overthrowing any existing order" (145–6). In short, street skating comprises "communal land use practices that challenge the primacy of market-rate private property . . . [and, in addition] the appropriation of found urban spaces through street skateboarding contests the given meanings of cities as growth machines" (Vivoni 2009, 146).

Another "perilous, graceful art" (Guss 2011, 73), parkour's antecedents are found in the *méthode naturelle*, developed by Georges Hébert, a physical educator, theorist, and instructor born in Paris in 1875. During military service in World War I, Hébert began to develop a set of physical practices based on the notion that to be strong was to be useful. He constituted these practices in contrast to what he saw as growing attachment to "competition and performance within contrived environmental spaces (such as a gymnasium)" and which he maintained had "negatively affected the corporeal and social development of youth" (Atkinson 2009, 171). Instead, Hébert designed three to six mile (five to ten kilometer) routes or courses known as *parcours* in wooded and other settings to encourage absolute gains in efficiency of mobility (Kidder 2012). Elemental in that work was training to enhance strength and speed by use of increasingly competently executed functional movements such as walking, running, jumping and landing, rolling, climbing, vaulting, and balancing, which remain fundamental to parkour (Edwardes 2009) and now typify many organized gym classes—a matter I raise again in chapter six. As important to Hébert as the physical fitness that derived from such discipline were its psychological and spiritual effects, among them the ability to overcome fear, marshal energy, exert willpower, and exhibit perseverance. Hébert's contention was that "by experiencing a variety of psycho-emotional states . . . during training, one cultivated a self-assurance that would lead to inner-peace" (Atkinson 2009, 171).

Among the beneficiaries of Hébert's methods was a soldier named Raymond Belle, who continued to promote the system, including to his sons. One, David, lived with his mother in Lisses, a commune in the region of Île-de-France, and there, with several other young men, he is said to have created *l'art du deplacement* or parkour, a term Belle attributes to his friend Hubert Kounde (Edwardes 2010). The influence of martial arts and "the acrobatic antics found in Jackie Chan movies" is palpable; so too, in training videos and other media, "the spirit of physicality and functionality prevalent in many ancient cultures and older disciplines" (Edwardes 2010, n.p.). Members later developed different forms of parkour following an apparent split in the group, with some, such as Sebastien Foucan, going on to champion freerunning, a practice like parkour embellished with more gymnastic flourishes, perhaps most famously captured in the Bond film *Casino Royale*. According to Atkinson (2009), in the BBC documentary entitled *Jump London*, Foucan described this foundational period in Lisses as one in which he and his companions looked at and thought about place in the ways children might, which suggests that parkour is both a focused discipline aiming to produce resolutely moral and fit young people, and playful resistance to normalizing spatial relations that arise at varied scales.

Suggesting a geopolitical relation between parkour and its place of advent, Nathan Guss (2011) describes Lisses as a product of suburban development in France in the period from around 1947 to around 1973, decades known as the *trente glorieuses*, elsewhere described as the three decades of the long boom from the 1950s. Such suburbs were typified by inexpensively designed and constructed mass housing, and over time have been subject to waves of in-migration by different groups, increasingly from non-Francophone backgrounds. Guss notes that many first generation *traceurs* were the offspring of first generation migrants and, in examining why parkour emerged in Paris and in France, considers its political salience. He posits that French urban policy had created spaces of constraint and disempowerment, and argues that parkour became an art by which to claim new commons—that is, "a shared public space and a collaborative intellectual or artistic endeavour" (83) and one that maps onto the varied but overlapping political goals of the Situationists, de Certeau, Lefebvre, and Foucault. Thus, in relation to the genealogy of skating and parkour as modes of disciplined conduct and playful mobility practiced by young people, recall from chapter one my observations that rhythmanalysis requires multiple senses through which to observe how life is weighted with power. Arguably, this 'weight' is experienced in the body, which is given effect via diverse anatomo politics, biopolitics, and governmental regimes. The body is also implicated in the constitution of heterotopia and heterochrony, and in the production of differing truths about what it means to be in suburbs and cities, not in the least as skaters and *traceurs*. Let me now elaborate on these points by reference to particular artifacts produced by some of their prominently public figures.

'Skate with Your Mentality'[4]

Skateboarding Explained is a seventy-minute "instructional DVD" presented by US-based professional skater Dan MacFarlane (2006) and produced by a consortium including Mentality Skateboards, Skateboarder Magazine, Ollie Pop bubble-gum company, and Lake Owen Camp—an overnight camp in Wisconsin specializing in gymnastics, BMX bike riding, inline skating, and skateboarding. The DVD features fundamentals, street and ramp techniques, and the most wanted 'tricks'. These sections follow a set of written warnings about the risks inherent in skateboarding and a brief introduction in which MacFarlane describes the anatomy of a skateboard: the board itself, with nose and tail, the rough upper surface known as the grip tape, bolts, trucks or metal axles, wheels, and bearings. Avoiding reference to any skateboard brand, manufacturer, or retail chain, MacFarlane suggests that skaters talk with providers about quality, price, and maintenance, the last of these to escape injury (on which see, for example, Lustenberger et al. 2010). Although MacFarlane does not provide any details on the range of safety gear for skaters, in all demonstrations throughout the DVD, in addition to Mentality T-shirts he wears a helmet.[5]

In the fundamentals section, on the understanding that unhurried mastery of the basics will make learning difficult tricks easy, MacFarlane underlines the importance of practice, discipline, balance, spatial awareness, and control. He suggests skaters find their stance by placing one foot on the board and pushing with the other, and then reversing the footing to see which feels right. Pushing with the right foot while the left is on the board is called 'goofy-footed' and the opposite is called 'regular-footed'. Standing on the board with feet perpendicular to its length where the right foot is near the nose bolts is also called 'goofy-footed'; when the left foot is near the nose bolts the stance is again called 'regular-footed'. Significant detail follows about how to manage body and board in concert, and MacFarlane makes the point that eventually the two work seamlessly. This comment constitutes skating as a socio-technical system about which, in general terms, Mimi Sheller and John Urry (2006, 209) ponder "what if we were to open up all sites, places, and materialities to the mobilities that are always already coursing through them?" In what becomes a rhythmic mantra about learning to "lean and look in the direction intended", MacFarlane's voice is interspersed with the sound of the wheels of his board on bitumen, wood, or metal as he describes and demonstrates how to place foot on board, bend knee, shift weight, push forward, gaze down, gaze back, gaze forward, swivel ankles, repeat, lean toward toes, lean towards heels, and so on.

Take, for example, a fundamental tail move. Place front foot near nose, and back foot on tail, look to back, press down on tail as you release nose pressure then push back on nose to bring front of board back, and practice small moves to pulse three hundred and sixty degrees before attempting a one hundred and eighty degree single turn. Turn your shoulders in the direction you intend to move, and gaze in the direction where you are about to go; momentum will follow. Equally detailed descriptions follow for 'tick-tacks' that provide a side-to-side momentum, 'switch stances' that enable use of both feet for different moves, and a range of other maneuvers with names such as 'fakie front-side' or 'backside 180' 'kick-turns'. Time and again, MacFarlane emphasizes the importance of visualization and positive thinking, of projecting oneself into a move, of learning on the flat before attempting tricks on an incline, of looking where you want to be because your body wants to go where you look and stays balanced, and of learning to fall and get up again.

Attention then turns to street and ramp skating, using the fundamentals and adding the 'ollie', a maneuver at the root of most street tricks. Ollies and their variants enable airborne movement (Figure 4.3). They assist skaters to mount the metal coping that typically surrounds ramps and bowls, as well as benches, handrails, and other street fixtures, positioning the board so it is parallel to the object being skated on, perpendicular to it, or perched by nose or tail on an edge to allow the skater to then 'drop in' or down to a larger skating surface, or 'stall'. Once mastered, these techniques enable skaters to progress to composite moves that involve rotating a board on short or long axes with skilled manipulations of feet and ankles, and which include backside and frontside 'pop shove-its', various forms of 'kick-flip', and 'heel-flips'.[6]

Other moves, such as 'front-side 50–50 grinds' or 'back-side nose slides' also require embodied knowledge of compound techniques. Of the back-side nose slide, for example, MacFarlane coaches skaters to learn to approach an obstacle, then ollie to position the nose of the board onto the obstacle's edge, and then grind along the edge balanced on the nose before setting down and stopping or continuing to move. The installations on which MacFarlane skates are part of Lake Owen infrastructure, the film understandably avoiding any shots of street skating, in general, which is heavily regulated in many settlements (Carr 2010; Chiu 2009; Stratford and Harwood 2001), and of

Figure 4.3   Executing an ollie. Source: Marco Prati, 2014 [iStock Photographs].

more controversial tricks, such as the 'coffin', which involves lying on the board on one's back while moving.

Skateboarding Explained offers many insights into the rhythms of skating. MacFarlane's approach is strongly reminiscent of Lefebvre's (2004) ideas about rhythm as simultaneously repeated and different, mechanical and organic, and discovered and created. At the same time, MacFarlane advocates disciplined practice, and this constitutes the body as both fact and project, "an existential condition and basis for our understanding of ourselves and our environment . . . [and] a phenomenon in the process of becoming" (Simonsen 2003, 164). Although his instructions make clear the critical importance of the gaze to skating, a moderately upright posture, the precise rotation of shoulders, or the strength of the core, inevitably MacFarlane's chief focus is on the feet and with the ways in which they mediate the project of becoming a skater.

Feet fascinated Lefebvre (2004, 28) when, from his Parisian apartment window, he observed that "people produce completely different noises when the cars stop: feet and words. From right to left and back again. And on the pavements along the perpendicular street . . . The harmony between what one sees and what one hears (from the window) is remarkable". Feet also intrigued de Certeau (1984, 97), for whom an understanding of the everyday practices of ordinary people begins "on ground level, with footsteps. They are myriad, but do not compose a series. They cannot be counted because each unit has a qualitative character: a style of tactile apprehension and kinaesthetic appropriation . . . They are not localized; it is rather they that spatialize". In turn, Ingold (2004, 332) suggests that, like hands, feet "mediate a historical engagement of the human organism, in its entirety, with the world around it", their motion rhythmically beating out "a pattern of lived time and space". Ingold refers to the expectation by planners and municipal authorities that movement by feet in cities is to leave no trace: no walking on the grass, no defacing of green or heritage surfaces, no marking of one's presence. In contrast to such expectation, skating leaves the marks of those who grind and kickflip and ollie as part of playful engagement with the built environment.

Foundational analysis of skating by Iain Borden (2001) is useful in relation to these ideas, Borden arguing by reference to Lefebvre's emancipatory agenda that skating challenges the functionalism and profound constraint of urban life, and asking how the city might be differently enunciated. Indeed, among the most worthy elements of his work are questions about what skateboarding reveals about cities and architecture, and their tendencies either to shut out or to valorize particular and narrowly circumscribed use values. Others have advanced Borden's work. For Nicholas Burrell (2012), the traces that skaters leave also mark a reappropriation of spaces from which they are often marginalized. These traces constitute a geopolitical sign, a spatialized vocabulary that speaks to other potential skaters and that may also alert authorities to new domains for intervention. For Rocco Castiglione (2012, 4), the perpetual dynamic of discovery and constraint between skaters and authorities is such that the latter fail to embrace skaters' "unique and intuitive ability to develop and design space" (Figure 4.4).

Castiglione then lists several features of the urban environment that skaters 'read' in ludic fashion, among them benches, bridges, gutters, lawns, pavement, parking lots, picnic tables, planks, planter boxes, playgrounds, ramps, roads, schoolyards, seawalls, sewer pipes, shopping malls, sidewalks, slides,

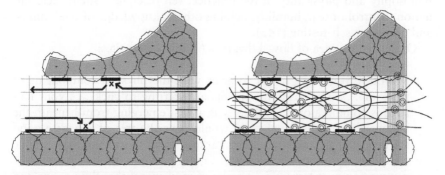

Figure 4.4   Bench diagram courtesy of Rocco Castiglione (2012, 37–8). Reproduced with permission. The image shows differential movement in the same space, the one linear, economical, and premised simply on moving through or sitting down (L), the other rhizomatic, playful, unpredictable—and hence less governable (R).

stairs, statues, street signs, and traffic. In various architectural schematics, he seeks to demonstrate how skating and other forms of movement can be accommodated simultaneously or, as I would put it, generously. The figuration of skateboarder movements that Castiglione captures includes much that has concerned me here: ludic geographies that manifest as *dérive* and *détournement*; mobilities that challenge normalized discourses and practices wrapped up in the capitalist city; and hints at the prospect that design can better accommodate all kinds of 'life between buildings' (Gehl Architects 2004; Gehl 2011).

"Be Focused, Silent, Give Your Heart"[7]

*Traceur*s and *traceuse*s engage in patterned movements through varied landscapes, seeking to question certain interests that "rhizomatically creep into the design, function, and representation of all social space and direct all cultural movement" (Atkinson 2009, 175). Yet there is a clear understanding among them that *l'art du deplacement* in fact should leave no trace. *Once Is Never: Training with Parkour Generations* is a film by Julie Angel (2009) and Parkour Generations[8] that illuminates this and other fundamental principles and practices of parkour (see also Angel 2011). Although it is not baldly instructional in the same sense as MacFarlane's (2006) work, *Once is Never* still focuses on key techniques and is augmented by other products such as Dan Edwardes's (2009) *The Parkour and Freerunning Handbook*. Richly illustrated with line drawings and photographs, the handbook describes the natural method of parkour (to distinguish an originary practice from derivative or breakaway forms); discusses holistic health and how to warm up, warm down, be safe, land, roll, jump, mount, vault, balance, complete wall runs, clear obstacles, drop, step, and swing and hang; and delineates advanced moves such as tic-tacs, crane jumps, kong vaults, and cat leaps (Figure 4.5). The handbook also elaborates on the philosophy and physicality of the practice: self-discipline, focus, determination, control of fear, humility, helping others, knowledge of one's limits, and persistent self-testing (143).

Of the central idea of flow, Edwardes (2009, 138) writes:

> So you're beginning to get to grips with the physical components of parkour . . . The basic movements are starting to click . . . your balance is coming along . . . fluidity is at last emerging from beneath the awkwardness you thought you would never be free of. That's a good start . . . But this is only a start. Parkour is far more than these simple component parts or techniques . . . good parkour is found *in continuous motion over terrain* . . . a series of movements that flow as one and instinctively.

Once is Never exemplifies these ideas using devices that parallel Angel's (2011, n.p.) doctoral work on the cinematic and theoretical elements of

Figure 4.5   Chris—cat leap—Abbey Road. Photograph courtesy of Andy Day, 2012. Reproduced with permission.

the practice, in which she sought to move away from parkour as spectacle (in both its common and Situationist interpretations) and return to the idea of parkour as everyday practice and lifestyle. Lest it be thought that such a lifestyle is all well and good for young people, the London-based organization Parkour Dance has started classes for aged pensioners in Bermondsey, in which "giant leaps [are replaced by] . . . careful steps, ensuring physical and mental boundaries are pushed safely, without broken bones . . . in the pursuit of keeping the mature students active and limber" (Jenkins 2013, n.p.). In Hobart, Tasmania, at least one gym has instigated 'balance classes' for seniors that comprise a circuit of ten obstacles to be negotiated safely and continuously during a forty-minute session weekly, and several gyms have developed or adopted 'cross-fit' as a way of training, which is indebted to many of parkour's maneuvers.

Running at one hour and fifty-six minutes, *Once is Never* has four sections entitled "New York—Seminar in Central Park"; "Ohio—USA Memorial Weekend Seminar"; "Sarcelles—Training in One of the Birthplaces of the Art in France"; and "Rendezvous—the First Three International Gatherings in London". This last section is subject to closer analysis here on the basis that it includes women and younger boys and girls, and people from several ethnic groups, although nothing is made of this diversity in the film, and perhaps to do so would be counter-intuitive. Either way, the variety of people involved sends a message that parkour is accessible to more than just young men. Each rendezvous is a visual diary or journal. Ever-present is a soundtrack of contemporary music, and clearly audible are the voices of members of Parkour Generations and their associates from the Yamakazi[9] offering guidance and instruction to around one hundred and fifty participants. There is no narrator, no organizational device explicitly unifying any or all of the meetings, and there is no sense of whether we are witness to three contiguous sessions or others spaced over days, weeks, or months. Fittingly, between beginning and end, the story drifts.

The first two rendezvous are located at a gymnasium in London. Dozens of practitioners train: warming up, practicing basic and advanced moves either singly or in groups, repeating them over and over, then cooling down and stretching out. People are told to remain relentless in their pursuit of discipline; that it is an *obligation* to do so. Embedded in the playful daring, and a possible effect of increasing speed of 'jamming' captured on film, is something I can only describe as an underlying sense that parkour involves a preparedness to 'defend' in ways that do not characterize skating, and that might recall *parcours* as physical fitness with a military ancestry. Yet, Jeffrey Kidder (2012, 237) compares parkour and skating as transforming space designed for one purpose into space for other purposes: "Like skateboarders, therefore, *traceurs* are offering a lived critique of the built environment—asserting playful creativity in the dead spaces of the city". In his own eighteen-month ethnographic engagement with *traceurs* based in Chicago, Kidder found they did not move great distances; they travel, train, jam, then walk to other areas, train again or jam, and he did not witness any efforts to execute flow-runs from one part of the city to another—although that may have been because of skill levels or concern about legal ramifications. Kidder also emphasized the paradoxes of parkour: a dangerous activity involving calculated preparation and training; requiring immense strength, coordination, and agility; not pursued by stereotypical athletes; and not constituted as rebellious in the way skateboarding has been seen. One participant in Kidder's study described the parkour community as extremely respectful and possessing a common love of navigating urban environments in ways unintended by designers: nevertheless, it is the "individuals willing to take the biggest risks, and do so with grace and style, that garner the most respect among their peers" (239).

In *Once is Never*, the third rendezvous is filmed outside Queen Elizabeth Hall on London's Southbank on a rainy day close enough to Christmas that crowds of other young people get about in Santa suits. Members of Parkour Generations and the Yamakazi speak about the virtues of such weather, and of being adaptable in all kinds of conditions, and they are quick to emphasize the need for additional care in the wet. Angel then captures footage of the practitioners working in pounding rain and puddles, fiercely concentrating because of the slipperiness of the built environment, continuing despite the relative chill of their bodies and consequent constraints on fluidity of movement, and enjoying the novelty—for some—of being out of the safe confines of the gymnasium. At one point, an instructor is caught patiently talking a young man through a jump from one bench to another—"I know you can do that . . . hey come on guys, help that guy . . . we start together, we finish together" (21 min 20 sec)—and then is heard celebrating his success. Later, several practitioners are filmed squatting against a wall, their task to stay in place for twenty-five minutes to build endurance. Such work is "training the spirit as much as the body . . . and after you've pushed through it, it's much easier" (25 min 40 sec). Describing his understanding of the history of parkour, Edwardes (2010, n.p.) also turns to this relationship between endurance and practice—forms of discipline and dressage that fold back to questions of conduct and the desire to flourish. In tones that, to me, are reminiscent of Aristotle's concerns with *eudaimonia*, Edwardes writes:

> The experienced *traceur*s do not engage in the activity in order to experience the adrenaline-rush that comes from excessive risk-taking. Rather, they view their practice as challenging themselves to overcome limitation and the restraints of fear and inhibition. The goal of complete functionality within any situation is paramount, and in order to achieve this state both mentally and physically it is necessary to face and overcome the challenges that are part and parcel of parkour training. The training in parkour enables the practitioner to learn to manage risk and through exposure to challenging situations to become a better, safer person all round.

ON THE POSSIBILITY OF BEING GENEROUS

Generosity, the propensity to show kindness, is elemental in ideas about prosperity and flourishing. According to Abraham Goldberg, Kevin Leyden, and Thomas Scotto (2012), there is a strong positive correlation between happiness and location—where we dwell—and such connectedness is integral to that relationship. However, their work does not refer to children and young people beyond consideration of their transportation to, for example, childcare facilities—an observation that implies that the authors are focused on adults' perspectives. Undoubtedly, there are many instances of unhappiness among those aged over eighteen, not least those arising from long hours spent in work they feel disconnected from, or in no work, or in commuting, including by means of the travail of footslogging, as Ingold (2004) would have it, and that is a subject to which I turn in chapter five. Nevertheless, there is significant evidence to suggest that, as part of their journey through the life-course, young people too often are rendered unhappy, not least by their engagements in cities and towns and by a lack of generosity on the part of adults.

Granted, this assertion is not without risk for two particular reasons. First, in my observational work on skating, I have occasionally witnessed unthinking acts by skaters that have startled elderly people—and my community consultations with the former have made clear the perilous effects on the elderly of injuries such as broken hips caused by falls precipitated by being thrown off-balance. This concern is widely held, and encapsulated in letters in the *British Medical Journal* where, for example, JR Oakley (1978, 115), a doctor from the Children's Hospital in Sheffield, writes: "Skateboarders are aware . . . they may injure themselves, but they seem to have scant regard for fellow users of pedestrian byways. I hope that future studies will include data concerning injuries caused by skateboards to other pedestrians, particularly young children and the elderly". Second, I have also felt ambivalent about gouges and wax marks on heritage sandstone, and appreciate the frustrations municipal authorities have balancing complex and competing demands such as engaging different groups and managing infrastructure. Concern for the latter often seems to triumph. Thus, situations in which young people are being moved on by police or council enforcement officers for skating or tracing seem far less grave than their being subject to violence or exploitation (Pain 2003; Scott and Steinberg 2008). Yet as Anderson and Smith (2001) have established, social and spatial relations are lived *emotionally*, and 'merely' being told to go away is still a signature form of rejection. Either way, a developed sense of geographical imagination and the potential of our cities and towns for ludic engagements simply *insists* that life could be otherwise, and alongside skaters, *traceurs*, and some others, progressive architects and designers know this.

Finally, then, let me return to the idea of play and its critical importance for us all, and especially for young people. Certainly, and with specific reference to parkour, Rawlinson and Guaralda (2011) describe space as a playground. Their first argument is that play is needed in cities (and arguably all settlements) on the grounds that it generates health and well-being across all ages, enhancing sense of memory and place, and revealing "the hidden terrains of desire and fear which affect the shape of our cities" (19). Their second argument is in two parts: that play can and should be creatively designed in ways that do not have to disrupt private space or capital; and that those engaged in the spatial disciplines, architects not the least among them, can and should *reframe* the ways in which design is understood and executed, Castiglione's (2012) work at Figure 4.4 being one such example. At the same time, Rawlinson and Guaralda (2011) usefully remind us that play has been a feature of settlements for centuries and that, in equal measure, it has been subject to governing practices, such as an instance in 1314 when the Lord Mayor of London, Nicholas de Farndone, banned football matches in the streets. So conflict is not new; nor segregation, decentralization, isolation, marginalization, and removal; nor the signification of some forms of play as deviance or criminal conduct by use of rid acts, summary offences, and restrictions.

Rawlinson and Guaralda (2011, 20) point out that play activities "must also contend with commerce and the novation of public space management and development to privately run governing bodies"; such developments are part of the technologies of agency and performance that transform citizens to consumers. Notwithstanding, the authors point to a (partial) resurgence of play in the city, and to activities which "begin to communicate both a rebellion against the oppressing socio-spatial norms of the city, and a declaration of the creative ludic potential and the playing spirit of humankind" (21), wandering off-score, as Edensor (2010c) would have it. For Rawlinson and Guaralda, different possibilities exist: occasions to create other spaces, heterotopia; mechanisms to design cooperative relationships into the fabric of settlements such as integrating play into design at the procurement stage of developing public spaces; and identifying opportunities for richly and deliberately multisensory environs. As they point out, ultimately, the "design of frame and path elements can be made to outlast our normative expectations of their use, and continue to provide a ludic and normative utility into the future... [right down to] material finishes and structural specification" (22–3). What may be required is a willingness among those with formal juridical authority to be generous: to enable what Rawlinson and Guaralda call natural sacrificial layers on buildings and streetscapes on which play can safely occur and for which materials and structures are appropriate; and to approve the creation of spaces that "require security or privacy, but that are outside of the façade, [which] must clearly communicate that they are not fit for appropriation" (23). Properly, Rawlinson and Guaralda note that those who would grind and trace will be able to read these signifiers, for they can be among the most perceptive and imaginative readers of space and place.

And so I return to *The Nicomachean Ethics*, in which pleasure may be seen as vulgar, as enslaving people to passing tastes and fads, or it may be honorable and based in practical wisdom (Book I:5). No matter how hard I try, I cannot imagine Aristotle practicing or countenancing skating or parkour—and that is not the point—but he does contend that there are certain goods, intrinsic goods, that do not derive from "some common element answering to one Idea" (Book I:6). In other words, diversity is a good thing. In this light, consider again Tara Woodyer's (2012) eloquent argument that playful spatial relations inform "the cultivation of a mode of ethical generosity". Play, then, has the propensity to provide what Lottie Child seeks: mobilizing joyful and rhythmic experiences, moving in space, and creating new spatial relations and engagements with people and place. Might skating, tracing, and other ludic forms of rhythmic mobility invite us to consider how we attend to spaces and spatial relations, to scales of engagement, and to the modes of emplacement and displacement that we design (or fail to design) into our settlements or where we dwell? And if made manifest, what might such innovations reveal about our propensity to understand prosperity in ways other than those attached to commerce or use value?

NOTES

1. Evans's inference about the cyclical nature of being corresponds with Lefebvre's (2004) ideas about the importance of rhythm, repetition, and cycles, which corresponds in different ways to the characteristics of skating and parkour as practices, habits, and modes of training and discipline.
2. Thick play means something akin to thick description—one that embraces not just a behavior or practice, but its context, such that it becomes highly meaningful.
3. No evaluation of the comparative *worth* of either is offered; to do so would be to defeat the purpose I have. Nevertheless, in undertaking these readings, comparison necessarily creeps into the narrative I present.
4. This phrase is drawn from the Mentality Skateboards (2011) website and an embedded advertisement featuring MacFarlane encouraging skaters to use their creativity to playfully explore the possibilities of the spaces and places they encounter. In this advertisement, he wears no helmet.
5. In the aforementioned consultancy undertaken for the Tasmanian Government during the legalization of 'small wheeled devices' on Australian roads in 2001, I strongly counseled that the incoming Tasmanian law require skaters to use helmets—which had been legislated in South Australia—but that was deemed impractical (Tasmanian Parliament 2001). A number of serious injuries have occurred since that time, most recently in April 2013 (Tasmanian Police 2013). That the skaters involved in this latest incident were behaving in a manner that was highly risky is certain; that the system failed them and others involved in the incident is absolutely clear.
6. Likely because they are high risk, a range of other and more gymnastic moves are not covered by MacFarlane (see, for example, msk1416 2013).
7. This phrase appears around 7.24 minutes into "Rendezvous", one of four sections in the parkour documentary *Once is Never* described in this section.
8. Parkour Generations is a company directed by Dan Edwardes, Francois "Forrest" Mahop, and Stephane Vigroux, which provides services in business—including film and television, live performances and events, certification, equipment, tactical advice to security companies, and collaborations with artists—choreographers, for example (Parkour Generations 2013).
9. Parkour history has it that the Yamakazi were a group of nine young men in Lisses, Yann Hnautra, Chau Belle, David Belle, Laurent Piemontesi, Sébastien Foucan, Guylain N'Guba Boyeke, Charles Perriere, Malik Diouf, and William Belle, who collectively started the discipline. According to Hnautra, the term Yamakazi is from the Lingala language of the Republic of Congo, and means strong person (Daniels 2012).

5   Encountering the Circle Line

When members of the Situationist International *drifted* around Paris, according to Guy Debord (1956) they let go of other concerns, including their usual motives for moving and acting (such as commuting and working). In advanced economies such as Australia, the labor or workforce has been understood as those aged from around fifteen to sixty-five. It is legally possible to work from fourteen years and nine months of age, and retirement at sixty-five is no longer compulsory in many occupations. Adulthood is deemed eighteen, but of course some members of the workforce are minors, and others are elderly. I am less definitive about the span of ages that concerns me in this chapter than I have been in previous chapters, and it is a decision I am comfortable with. Certainly, among participants in Sandy Nicholson's (2011) *0 to 100 Project*, ideas of what constitutes adulthood do vary. Michael Leggieri, aged twenty-nine, revealed that it "feels like childhood's got a little bit more stretched out than [for] my parent's generation". Jeff Anderson, forty, described adulthood in terms of the symptoms of prosperity as Aristotle also might have: "career, house, wife, family, friends, all that kind of good stuff". At forty-seven, Lori Livingstone noted that as "I get older, the word 'old' keeps getting pushed out, so to me, old is over 70 now".

To drift, then, is to model differently the dynamics of human relationships—to displace, dislocate, and mis/reappropriate in order to see anew (Kofman and Lebas 1996). Thus, the Situationists' *dérive* and *détournement* stand in contrast to the daily grind of adult labor force activity. Australian painter John Brack captures this daily grind in *Collins St. 5 p.m.*, a rich visual commentary on peak-hour Melbourne in the 1950s (Figure 5.1). The work "depicts people emotionally closed down . . . despite their close proximity, suggesting a loss of individuality and a lack of social cohesion among the masses" (National Gallery of Victoria n.d., n.p.). *Ennui* prevails, but one can also imagine that many of those in the crowds will descend stairs to trains at the Flinders Street Station or ascend onto trams and buses and return to places wherein they dwell. That is, I think, a far cry from the sense of grave mistrust of one's fellows that seems to typify the rush hour of present times in so many places.

Figure 5.1   John Brack (Australia 1920–1999) *Collins St. 5 p.m.* 1955, oil on canvas, 114.8 × 162.8 cm. Courtesy of the National Gallery of Victoria, Melbourne. Purchased, 1956. © National Gallery of Victoria. Reproduced with permission.

Rush Hour is a YouTube film on parkour of less than two minutes produced by the British Broadcasting Corporation (2006), which features David Belle, aforementioned foundational figure of the movement. Seated in a modern office, Belle gazes at his desk on which stands a temple structure and martial arts figurine—signifying the idea of parkour as a discipline—and then glances at his watch. Removing his jacket, shirt, belt, and trousers to reveal sweat pants and a toned physique[1] Belle opens a window and progressively maneuvers himself up the multi-story window frame to the rooftop—facing *down* in plank formation, hands and feet moving *backwards up* the verticals of the frame. On the streets below are lines of cars, trucks, motorcycles, and pedestrians caught in the snarl of traffic. Witnessed with incredulity by a woman in an office nearby, Belle executes a wide-gripped handstand on the railing of a rooftop balcony, then disappears from her view, tracing across rooflines and fire escapes to his apartment. At one point, he stops to straighten an aerial he has disturbed, righting a boy's television reception in an apartment below—a reference to the idea that tracing should not harm. Our gaze aside, this journey home is constituted as unseen, a private daily practice—but certainly not routine: Belle's novel daily commute being utterly focused on running, vaulting, rolling, climbing, tic-tacking, flipping, spectacular leaping, and sliding—no two journeys are experienced in the same way.[2] The freedom and playfulness of his trace are emphasized by shots of the crush of people below him, and underlined by the dulcet tones of Dean Martin's (1953) *Sway*: "Other dancers may be on the floor / Dear, but my eyes will see only you / Only you have the magic technique / When we sway I go weak".

Oli Mould (2009, 741) suggests that Belle's movement challenges the rationality of urban grids and "reappropriates the urban built environment from a striated space to a more fluid smooth space". Such reappropriation infers an inverse co-constitutive relationship between lines—which are striated spaces, and points—which are smooth spaces between lines; the former close off, the latter mark nomadic distributions in open space. In other words, according to Mould (2009, 742), "walls, rails, staircases, car parks, or any other of the single objects of the built environment" are needed for "drawing your lines of flight"; they reveal the multiplicity of meanings of any single urban object and of many objects in assemblage. However, Mould departs from this reading when it requires a sense of flight as "enacting a violence against the striated city space" (742). Instead, he argues that *Rush Hour* presents flight as an idiom both *non-violent and non-reactionary*, and one committed to problem-solving. Assuredly, the *traceur* and the *traceuse* do call on their senses, read and experience the haptic geographies of their lines of flight, and create ludic pathways; and, furthermore, Belle deploys these technologies in the place of 'normal' rush hour mobility—indeed *as* a creatively reappropriative act. But *Rush Hour* evokes more: viewed enough times, it shifts from being spectacle and entertainment to become a statement about the power of intimate, repeated, embodied, and thoughtful connection to buildings and space—his dwelling-in-motion (Sheller and Urry 2006). Belle's productive counter-moves invite consideration of the questions *how do adults experience the practice of commuting?* And, *in what ways might commuting be considered dwelling-in-motion?* These are two of three questions that concern me in this chapter. Prompted by Mould's (2009) observations about non-violence and its opposite, the final question requires two points of explanation, and its articulation is delayed until those explanations are in place, as follows.

First, Mould's (2009) analysis brings to mind Lefebvre's (2004, 21) specific guidance on the demeanor of the rhythmanalyst: she or he must call on all the senses without privileging any; think with the body as a lived (and therefore spatialized) temporality; be garbed "in this tissue of the lived, of the everyday"; and grasp both the rhythm perceived and its whole context, arriving "at the concrete through experience". For Lefebvre, such efforts require retraining, so that the rhythmanalyst comes "to 'listen' to a house, a street, a town, as an audience listens to a symphony" (22).[3] I have sought to apply these guidelines by considering the geographies, mobilities, and rhythms of commuting as an adult, using the tools afforded by affective and somatic responses to experience (de Freitas 2011). Circumstance has allowed that I settle on the London Underground's Circle Line, parts of which are *the* originary subway commuter line. Each year that I attend the Royal Geographical Society and Institute of British Geographers conference at Kensington Gore, I spend a day on the Circle Line. At each stop, I ascend, stand in one position, and take a series of photographs around three hundred and sixty degrees, descend back to the platform for the next train, and repeat the process until I have covered all stations, taking occasional breaks above ground in coffee shops. All the while, I smell, taste, listen, touch, move, and experience the world with proprioception and kinesthesia.

Second, then, my encounters on one of the great commuter lines of the London Underground have revealed so many geographies, mobilities, and rhythms underlining how commuting reaches tendrils into daily life—including the ways in which we organize our most personal routines. Commuting influences, even structures, the pace at which people walk or ride (perhaps to a transit station or directly to work) or calculate the time of departure from homes (perhaps via coffee shops or schools) and thence onto and into arterial traffic flows (Jones, P. 2012; O'Dell 2009; Wener et al. 2003). Commuting is instrumental in the formation of transient and more lasting networks, engagements, and practices; in this sense, it might be read as intimate, habitual, ephemeral, and constant—the means by which we come to dwell and move simultaneously. In this vein, Tom O'Dell (2009) describes how commuters make places to dwell in train compartments, colonizing the space of the train using spatial, auditory, emotive, and other tactics—their intent to be safe and comfortable. O'Dell notes, for example, how commuting creates a "world of micro-calculations . . . which come to be highly meaningful when woven together in the pulse of daily life. And this is not an insignificant insight . . . [since it is in such spaces and flows] where the drama of daily life is anchored, and understandings of it begin to emerge" (96). All the while, commuting 'acquires' things; it connects (and connects *with*) multiple other sites tied into multiple other rhythms, mobilities, and geographies (Figure 5.2): people, dogs, strollers, turnstiles, escalators, elevators, stairs, and platforms; ticket, entertainment, and tourist booths; newsstands and information stands; washrooms, eateries, and chain stores; art works and poetry; offices and monuments.

The bottom line is that each time I complete the Circle Line journey, and without any evidence of the validity of my concern, I have a lingering sense of anxiety about being under surveillance and, foolish though it seems, end up a little worried about being mistaken for someone with questionable intentions. Such unease prompts me to ask a third question here: *what does the terror of violent disruptions to commuting as dwelling-in-motion mean for how we think about conduct and flourishing?* Thus, I want to explore the vulnerabilities of commuting as dwelling-in-motion because those auto-ethnographic moments, originally chosen by me to experiment with rhythmanalysis, have forced me to confront something else. The attacks in the USA on September 11, 2001, almost universally known as the events of 9/11, involved transport and commuters as well as travelers; so too did the

Figure 5.2   Ascension from the London Underground. Source: Stratford, 2011.

July 7, 2005 bombings in London, which are known as the 7/7 attacks; and so too did many others elsewhere. Elden (2009, xiv–xv), for example, maps both the locations of attacks by nonstate terrorists since 1998, and the states which have experienced interventions from the United States and Allies since the end of the Cold War. At the same time, Elden notes that tallies "risk losing sight—and losing the site—of the problem . . . [putting] accountancy in place of grief. Let us not forget, then, that these events are a political, spatial, and temporal marker; yet they are one that we give a particular significance to through our complicity in a construction . . . 9/11 masks the spatial context of the events in favor of a temporal indication".

Without doubt, then, the securitization of transportation and of life has escalated significantly; that much we all know. But beyond all the other ugliness and tragedy that is enfolded in such conduct, terror's power partly derives from the fact that it so profoundly disrupts—indeed it decimates—our capacity to dwell-in-motion and thus is an *intimate* attack on our capacity to flourish in our daily rhythms, mobilities, and geographies.

COMMUTING AS DWELLING-IN-MOTION

Commutation is a term denoting the processes involved in change; it stems from the Latin *commutare*, to change altogether, alter wholly, to exchange, or to interchange. The word once was used to describe certain parish or county roads which required those using it to pay a sum of money for the repair and upkeep of the path (OED). With the advent of the railway in the United States, the commutation ticket was developed so that its purchaser could travel repeatedly at reduced cost the same route over days, weeks, or seasons. In formal taxation systems, commuting is deemed non-deductible regular travel from primary places of residence to places of work or full-time study (Australian Government. Australian Taxation Office 2013; United States Government. Internal Revenue Service 2012). In common parlance, commuting is the grind of billions of people daily, a significant proportion of them adults engaged in labor force activities.[4] Originally a mostly bipedal affair, commuting was later enhanced by the advent of the omnibus, tram, and rail, and then by the technologies and cultural practices of automobility (on which see, for example, Böhm et al. 2006; Bonham 2006; Sheller 2004; Urry 1999, 2006). Certainly, among very large numbers of adults, the experience of commuting closely mirrors the angst, noise, congestion, pollution, and rush above which David Belle so fluidly moves, and which has been the subject of both popular humor and critical analysis. Many of us remember, for example, Disney cartoons of Goofy as the benign and civic Mr. Walker, a 1950s American male who lives in the suburbs and makes his way to the city each day by car along increasingly complicated road systems, in the process changing into his breathtakingly rude alter-ego, Mr. Wheeler. The moment he gets in his car, he becomes a monster in a metal cocoon. Such a tendency Deborah Lupton (1999, 70) describes as symptomatic of cyborg embodiments—assemblages of the human and technological—and of forms of rage that mark a sense of fear "that the world is changing too rapidly, that there are greater uncertainties about life, that there is a breakdown in traditional values and increase in incivility".

Commuting and the stresses that it engenders are associated with diminished sense of individuals' subjective well-being (Dolan, Peasgood, and White 2008). In one economic analysis of this effect, Stutzer and Frey (2008, 339) describe commuting as "an important aspect of our lives that demands a lot of our valuable time" and they suggest that most people experience commuting as physically and mentally burdensome. Noting that commuting is one among many decisions made daily by apparently 'rational' individuals, Stutzer and Frey also question traditional economic understandings that assume it produces a state of choice equilibrium between housing tenure and experience, on one hand, and labor market and work experience, on the other. Therefore, any stress incurred while commuting should be compensated for by either the domicile from which the commuter departs, or the financial or intrinsic rewards at the workplace to which she or he travels. However, drawing on a data set collected since 1991 by the German Socio-Economic Panel,[5] Stutzer and Frey reach other conclusions. They find that "people with long journeys to and from work are systematically worse off and report significantly lower subjective well-being. For economists, this result on commuting is paradoxical" (340). Two clarifications are advanced by them. The first clarification is that there may be cases where families or significant others associated with commuters benefit from their engagement in long journeys insofar as household well-being is equalized, even if individual well-being is not. The second is that people are often unable to extract themselves from commuting and all that attaches to it—rent and mortgages, and wages and salaries, for example.

Ultimately, Stutzer and Frey speculate about the limitations of economic explanations for commuters' sense of well-being and suggest, instead, the utility of behavioral understandings. For example, commuters "may make mistakes when they predict their adaptation to daily commuting stress [or may have] . . . limited self-control and insufficient energy" (361–2). Stutzer and Frey seem to suggest that commuters could increase subjective well-being by exacting greater self-control and investing more energy in how they organize, enact, think, and feel about commuting. Appreciating that people could adapt to commuting in ways that improve their sense of well-being, nevertheless, Stutzer's and Frey's work is silent on matters such as gender or sense of identity in place (belonging) that would confound such a simple solution, vested in prudential self-care. These observations recall Dean's (1999) thoughts on the ways in which self-governing and the governing of others give effect to different ways of producing truth.

Gender differentials in commuting are noteworthy. Jennifer Roberts, Robert Hodgson, and Paul Dolan (2011) suggest that researchers are reasonably confident about the relationships between physical health, status in the labor market, and well-being. Highlighting the relative lack of studies on commuting and well-being, they observe that the 'average' British commuter traveled fifty-four minutes a day in 2006, which is an increase of six minutes over the period since 1997, and an upward trend common across Europe and the United States (Stutzer and Frey 2008, 342, Figure 1). More recent reports suggest that many in the UK are traveling for longer—as much as three hours each way—on each working day (Gregor 2013). Among key stressors in commuting are the time taken to travel, the unpredictability of human and non-human others, a perceived loss of control over one's fate, and boredom. Women tend to work shorter and less flexible hours in formally paid employment, and longer hours in informal and unpaid roles. They have lower wages, on average, and are more likely to work in occupations deemed of lower status that often can be sourced in local labor markets close to home. It might seem reasonable to assume, at first glance, that women will commute less often, and less far, than men—but even were that not contestable, it would not be reasonable to assume that their trips are less complicated or less demanding given the multiple tasks and many scales of engagement involved in social reproduction, for which they retain primary responsibility. Thus, even accounting for compensation, such as was described above, analyses produced using British Household Panel Survey[6] data suggest that women's "sensitivity to commuting time seems to be a result of their larger responsibility for day-to-day household tasks, including childcare and housework. Only those men with pre-school age children appear to be adversely affected by commuting, and even then the effect is smaller than for women with or without children" (Roberts et al. 2011, 1065).

Some studies point to the possibility that stress in commuting implicates affective processes of bodily transformation—for example, having consciously to stay awake and alert during long, early-morning trips (Bissell 2014a). Others suggest that stress may be exacerbated by having to negotiate multiple places of belonging—moorings as well as movements. According to Gil Viry, Vincent Kaufmann, and Eric Widmer (2009, 122–3), commuters are faced with occupying "poly-places [and are] . . . forced to develop and maintain social anchorings [family and friends] in different places, sometimes far apart from each other". Nicola Hilti (2009, 145) also notes the rise in incidence of multilocal living, which she understands as comprising "a way of organising everyday life in and between different homes" requiring residents to live in one place and work in another, which may be at some distance and require a second domicile. Rather than ascribing positive or negative attributes or outcomes to such arrangements, Hilti views multilocal living as "the structuring framework (of opportunities and constraints) for everyday life and . . . an everyday practice" (145). Her approach avoids the polarity attending pronouncements that commuting is *necessarily* and *inevitably* stressful, and that solutions to any angst experienced reside *necessarily* and *inevitably* with the commuter alone. It also recognizes that mobility and multilocal living are entwined, the latter an expression of the need or desire to move and settle. Reflecting on her studies of commuting journeys made by multilocalists, Hilti moves between examining commuting as productive and inequitable and stressful. On one hand, she refers to common practices among commuters to use the gap that is travel time in order to sleep, rest, relax, learn, read, prepare their thoughts and strategies, interact with dependants in tow, and experience simply being. This kind of practice does, indeed, enliven Casey's (1996) ideas about place as an event, noted in chapter one.

On the other hand, Hilti acknowledges that commuting can be stressful because it is entangled in inequitable systems—for example, those pertaining to employment and housing. She notes, too, the existence of markets catering to commuter comfort and infrastructure in which inequalities are deliberately inscribed into social processes and spatial practices: witness classes of seating, the availability of sockets for laptops, or luggage room in public transit. Notwithstanding, multilocal living and commuting are often understood as providing people with the means to enhance their socioeconomic status, since the jobs involved often attract high wages; their consequent spending in more than one location may also even out structural differences in regional economies. Hilti argues that against "this backdrop *the in-between, the interstices*, are identified as crucial for studying . . . multilocal living . . . Interstices have a specific meaning in the context of a multilocal organisation of everyday life" (146; original emphasis). For Hilti, attention to the in-between is a key means to disrupt the binary invoked by thinking of settlement as stasis, and mobility as disruption.

Others work to similar ends. Edensor (2010a, 189) has written on commuting's geographies, mobilities, and rhythms to unsettle its representation as "a dystopian, alienating practice" and rethink its capacity to offer "a rich variety of pleasures and frustrations". Commuting, he reminds us, produces many rhythms (personal schedules, train timetables), mobilities (by shoe, board, wheel, track, or wing; at a snail's pace or speedily) and geographies (concentric, radial, and other patterns of land use, for example). Among the insights he derives is an apprehension that places are being formed and reformed constantly in contingent and ephemeral ways by means of connections and interrelationships—networked mobilities involving "capital, persons, objects, signs and information" (190). Places are always also dynamically in process, their spatial arrangements shifting and changing, often in rhythm with the mobile elements that flow across them. Nevertheless, there is a *sense* that place and the things that comprise it are relatively stable. Edensor suggests that this apprehension of spatial order in place is maintained, in particular in transport networks, by the "organised braiding of multiple mobile rhythms" (190) that constitute and are constituted by traffic management systems, travel practices, forms of transportation, and "the characteristics of the space moved through". This idea of moving through leads Edensor to consider the ways in which the experience of space and place are shaped by mobility and the rhythms that accompany it. He refers to the "speed, pace, and periodicity of a journey" and their collective capacity to give effect to "a stretched out, linear apprehension" (191) such as might be formed on road or rail. From such apprehensions emerges a particularly important *mode of being*: a "material and sociable dwelling-in-motion" (Sheller and Urry 2006, 214).

Ole Jensen (2009, 139) takes up this idea of dwelling in order to move discussion beyond the binary implied by ontologies of sedentarism and nomadism; his proposition is that dwelling is integral to mobility, as evidenced by the relations between the "fixed and bounded sites" that are *enclaves* and the "infrastructure channels and transit spaces" that form *armatures*. Jensen offers up conceptual spaces in which to reconsider commuting as creative and to view armatures as more than "generic non-place[s]" (140) that merely convey people from A to B and erase any sense of connection, engagement, community, or cultural significance. Such ideas are paralleled by the possibility of imagining "bodies that develop particular aptitudes for movement . . . [and become] . . . an effect carved out of duration . . . habit becomes the driver" (Bissell in Adey et al. 2012, 186). According to Jensen (2009, 145), the armature may be constituted as a "space of interaction and meaning—a site of cultural expression and performance". So conceived, the armature enables moments of encounter among mobile subjects engaged in molecular politics using a range of strategies and tactics to create the public domain, and to inform debates about the good city. In short, Jensen argues that urban travel should and can be pleasurable, productive of a "new aesthetics of mobility" (153) and of other spaces, heterotopia. Jensen's primary concern, then, is with armatures and the reconstitution of the public domain, and his work represents an important contribution to theorizations beyond sedentarism and nomadism.

Likewise, Phillip Vannini (2011) seeks to challenge this binary between the sedentary and nomadic—in this instance by reference to the ferry lineup. He invites mobilities scholars to work beyond a metropolitan focus and, in ways that broadly parallel Jensen (2009), argues for an understanding of the ferry lineup as an ephemeral mooring in which all sorts of relational performances manifest. Drawing on data gathered over three years plying the British Columbia ferry system, Vannini (2011) persuasively calls for a reconceptualization of the lineup as a multifaceted space in which people read, sleep, think, pace, and engage with each other. Employing a musical term, *tempo rubato* (stolen time), Vannini observes how those who travel on the ferries, and thus queue for their opportunity to board them, use time in tactical and productive ways.

Thus, among other things, commuting is a rhythmic form of dwelling-in-motion. As Edensor (2010a, 192) would have it, commuting "sews places together". Among the first technologies to enable this stitching was the railroad. In developing his understanding of rhythmanalysis and its spatialities, Edensor makes numerous points about rail's affordances—qualities such as linearity, smoothness, network, or capacity to engender among passengers experiences that can be styled as "regulated synchronicity" (195). His own account values the different rhythms and mobilities of rail travel, unsettling other narratives, such as that advanced by de Certeau (1984) of the rail traveler as incarcerated, unchanging, pigeonholed, and numbered; the restroom a singular space in which to escape an oppressive regimentation. Yet, early in its history, railroad travel promised to shrink space and time and offer immense freedoms. For at least one early commentator writing in the *Quarterly Review* in 1839, and quoted by Wolfgang Schivelbusch (1986, 34), this prospect might also greatly increase people's capacity to dwell together: thus, if "railroads, even at our present simmering rate of travelling, were to be suddenly established all over England, the whole population of the country would, speaking metaphorically, at once advance *en masse*, and place their chairs nearer to the fireside of their metropolis by two-thirds of the time which now separates them from it". Indeed, the acceleration of transport would then increase the rate at which people would draw near to each other such that "all would proportionally approach the national hearth. As distances were thus annihilated, the surface of our country would, as it were, shrivel in size until it became not much bigger than one immense city" (34).

It is not lost on some that such ideas fail to account for other effects of rail—among them disruptions to the particularity of place and the singularity of times that attached to individual places. Cresswell (2006), for example, recognizes that the development of rail incurred social and economic injustices, the loss of place, and the disappearance of local time zones; nevertheless, he acknowledges that rail enabled millions to travel who otherwise would not. Adey (2010b, 2) notes of his own train commute from Liverpool to Keele: "In my relatively immobile state on the train, I am able to work, receive text messages, edit this manuscript. At other times I just sit back and try to relax". This description of productivity and self-motivated engagement is mirrored in extensive studies of train commuting and time use that involved surveying tens of thousands of UK rail passengers in 2004 and 2010 (Lyons, Jain and Holley 2007; Lyons et al. 2013). These observations and studies also echo David Neft's (1959) argument that transportation routes and schedules in London, Paris, and New York are designed by reference to the habits of commuters, and in turn then decisively influence those habits—some of which we can now read as dwelling-in-motion. In all cases, Neft found that commuters and commuting technologies were mutually affected by the extent and configuration of the networks onto which they mapped. For him, London was the standout in terms of the number of intersecting lines that enable the high levels of connectivity[7] that were anticipated by our commentator from 1839. It is to that city and its rail system that attention now turns.

THE LONDON UNDERGROUND

According to Christian Wolmar (2004), by the mid-nineteenth century London's roads were sclerotic, the city chaotic and imperiled by a combination of internationally fueled affluence and rural to urban migration that propelled up to a quarter of a million people into the city to labor each day. Work on *The Making of Modern London by* Gavin Weightman et al. (2007) includes an extensive account of both the city and its emergent rail networks. The point is made that walking was a long-standing and widespread mode of commuting from suburbs proximate to the inner city, often requiring workers to pay tolls to use privately owned bridges spanning the Thames. In contrast, those who were affluent and enjoyed more flexible working hours lived further out and could afford carriages or long distance coach fares. Despite the relatively rapid pace of technological change, alongside walking it was horse-drawn transport that remained conspicuous until the early twentieth century. Certainly, for decades the influence of the horse omnibus was more significant than that of the railway. Pioneered in the late 1820s by George Shillibeer, the horse omnibus faced initial challenges in terms of the width and congestion of roads, or the effects of monopolies on the use of certain roads by those operating hansom cabs. Shillibeer's own ventures were limited in terms of return on investment. Nevertheless, the worth of his omnibus design and the capacity of the vehicle to carry twenty-four people meant that others were finally able to challenge the cab monopoly in the latter 1830s. As a result, the omnibus became commonplace over the next three decades, its later demise precipitated by what Weightman and his colleagues drolly call a fuel shortage—successive poor harvests of hay and oats—and decisions made about urban governance and planning that favored other modes of transport.

The growth and influence of railway lines stem from a convergence of extraordinary wealth and entrepreneurialism, employment, land specula-tion and housing development, and innovations in transport and logistics. They are archetypally creative armatures, to use Jensen's (2009) aforemen-tioned framework. Indeed, railways "excited the imagination of the age . . . Like all other developments at the time, they were established and run by individual commercial companies and their activities were for the most part only haphazardly controlled by Parliament. It was only gradually that their impact on London was understood" (Weightman et al. 2007, 57). The first line between Deptford and Spa Road in south London's Bermondsey district in the borough of Southwark was started in February 1836, and reached the southern end of London Bridge the following December. Its construction established a pattern of so-called improvements repeated many times—the displacement of the most marginal city dwellers and the leveling of slum res-idences on which they depended; another instantiation of 'moving on' that involves profound forms of rejection. Indeed, Weightman et al. (2007, 58) underscore the point that it was "easier for the railways to get close to the centre from the south, where the land was not so heavily built up, and where there were fewer of the wealthier classes to resist the invasion". For decades, that powerful resistance was palpable, with all main-line stations distanced from the affluent West End—Euston in 1837, King's Cross in 1852, and Paddington in 1854—and with only one other station opened in 1841 in the city at Fenchurch Street on the London and Blackwell line near Tower Hill.

Then, in the 1860s, there was a new push in railway building. Wolmar points out that by that time, British entrepreneurs had awoken to the pos-sibilities of passenger rail internal to the city, and not simply those lines between city and country that privileged the movement of freight. The proving ground for passenger transport was, Wolmar suggests, the London and Greenwich line, which opened in 1836 and carried one thousand five hundred passengers a day for fares ranging from half a shilling in open cars to a shilling in imperial carriages. As more rail lines came into com-mission, and as more people took advantage of commutation tickets, the rail entrepreneurs and shareholders came to realize the commercial value of commuting and other benefits such as leisure travel and the attendant need for accommodation. Weightman and his colleagues suggest that the most palpable symbol of the boosterism and optimism engendered by the railway is perhaps St Pancras Station, with its luxury hotel, adjacent to the British Library. It is here that I always begin my encounter with the Circle Line, starting out from the base of Eduardo Paolozzi's 1995 sculpture 'Newton'

Figure 5.3 Eduardo Paolozzi (1924–2005) 'Newton', *After William Blake*, 1995, bronze sculpture, British Library. Source: Stratford, 2009.

After William Blake (Figure 5.3).[8] Given Newton's ideas on the laws of motion, it has always seemed apt to start here.

Nevertheless, the success of the rail system was countered by growing discontentment about its significant diseconomies and social injustices, and made public and political by the likes of Henry Mayhew, playwright, journalist, co-founder of the satirical journal *Punch*, and author of *London Labour and the London Poor*. In fact, the railway companies "were gradually forced to change their policy by legislation giving them the responsibility of doing something for the people they made homeless by the building of new stations" (Weightman et al. 2007, 64). Parliament began to place caveats on approvals for demolitions—the directors of the North London line were compelled to agree that they would provide services at low fares for workmen in order to gain the government's permission to build the Broad Street Station in 1865. Then, in 1883, Parliament passed the Cheap Trains Act, which made railway companies responsible for providing 'workmen's trains' at certain times of day. Little wonder that ideas for a system of underground rail lines eventually fell on fertile ground—after no less than fifty-five proposals for similar schemes had been submitted to the Private Bill office of the House of Commons in 1863 alone.

According to David Welsh (2010, 20), it was Charles Pearson's "plan for 'Suburban Residences for London Mechanics' that turned out to be the blueprint for the Metropolitan railway". It was the Metropolitan that led to the realization of the London Underground and the burgeoning suburbs of northwest London—which came to be known as Metroland. Welsh argues that at least some of "the ideological roots of the Metropolitan Railway were to be found in a reforming utopian impulse" (21), an age-defining urge to build the infrastructure for people to prosper and flourish. If one considers that this desire was guided by Pearson from the mid-1840s, it is relatively straightforward to understand its success. In the *Oxford Dictionary of National Biography* it is noted that Pearson was City solicitor from 1839 until his death in 1862, and Member of Parliament for Lambeth between 1847 and 1850 (Robbins 2004). He is described as a tireless campaigner for social justice, public health, free trade, the abolition of capital punishment, reforms to the treatment of juvenile offenders, and planning for settlements and transport. Such zeal is captured in *Proceedings at a Public Meeting Held at the London Tavern on the 1ˢᵗ December, 1858*. In this text, Pearson's passion for planning is tangible; so too is his (obviously contextual) concern with others' well-being:

> The noble street improvements undertaken by the Corporation [of London] have swept and are about to sweep away thousands of industrious artizans [*sic*] and mechanics from their humble dwellings, to make way for the spacious streets and splendid warehouses destined in this age of progress to take their place . . . My project [rather than one proposed by the Corporation that is too expensive for those displaced] will immediately apply a partial, and ultimately a perfect remedy for the overcrowding of the habitations of the poor, by giving to a portion of the industrious and deserving, the incalculable blessing of suburban residence, by means of cheap and frequent trains to carry the head of the family daily to and from his place of work. (Pearson 1859, 38)

Thus, in 1863, the world's first underground railway began between Paddington and Farringdon Street. Its northwesterly route from Paddington was to become the Metropolitan Line. From an easterly extension to Moorgate in 1865, and south and then east from Paddington, services were later extended in both directions, and ultimately formed the Circle Line. Wolmar describes the Circle Line as transformative. It is also a fascinating exemplar of tensions in the differential interpretation of flourishing and the pull of specific and narrowly-privatized ideas about prosperity that seem to run counter to Pearson's utopian ideals. The Line was the first network of underground routes that were all located beneath the city—some by dint of cut and cover methods, some by deep tunneling—and that served stations local to the inner city. It was also the first line developed in the latter decades of the nineteenth century that, according to Wolmar (2004, 70), was built by "government diktat", informed by successive Royal Commissions, and requiring full funding from the private sector. It resulted in a measure of *detente* between the oppositional chief operatives of the Metropolitan and District line companies. Certainly, over the 1870s and 1880s, the two were forced to cooperate to safeguard the line's completion when a group of City of London financiers (who had formed the Metropolitan Inner Circle Completion Company to ensure just that end) failed to raise sufficient capital to achieve their goals. When the Circle Line was complete, the Metropolitan owned seven of the fifteen route miles (eleven to seventeen kilometers) and ran clockwise around the outer track. Running in the opposite direction on the inside track, the District owned just under six route miles (just under ten kilometers). Jointly owned were just over two route miles (three kilometers). As Wolmar (2004, 86) observes, "While this could, with goodwill and sensible management, have been an effective way of operating the seventy- to eighty-minute round trip covering [then] twenty-seven stations, the hostility and antipathy between the two ensure that it was a recipe for chaos". Added to this enmity, the route had been chosen by a Parliamentary Commission, not by a market analysis, and, as Wolmar (2004, 89) reports it, "was not very useful until a broader network of Underground lines was connected with it". Thus, despite servicing over one hundred and fourteen million passenger journeys in its first year, and generating £1,012,000 in ticket revenue, the venture gave shareholders few dividends.

Later, the Philadelphia-born magnate Charles Tyson Yerkes acquired the Circle and District lines, along with other systems experiencing faltering economic viability. Wolmar (2004, 163) describes Yerkes's enterprises as a tangled "network of construction companies, operating companies and holding companies of interlocking directorships and friendly contracts, of financial manipulation and political corruption". It was Yerkes who founded the Underground Electric Railways Company of London Limited (UERL), which dominated the London transport network until the creation of London Transport in 1933. At the same time, much of the whole system's success in modern times is attributable to the work of Albert Stanley (later Lord Ashfield). Recruited to the UERL in 1901, Stanley created a vision for an integrated transport system for London, which saw his rise to the chairmanship of London Transport. In tandem with Frank Pick, who became London Transport's chief of staff, over decades, Stanley championed a harmonized aesthetic to the appearance of stations, ticketing reforms, logos, signage, timetabling innovations, and a corporate commitment to reduced journey times and limited delays—a series of innovations that have become known as the passenger experience.

JULY 7, 2005

Stanley and Pick's promotion of systemized publicity for London Transport routes anticipated the London Underground network diagram—the Tube Map—made famous by Harry Beck in the 1930s (Vertesi 2008). Of this masterful codification of space, David Ashford (2013, 64) notes that the Tube map is more than abstract space, and invites ready access to a system of transport that promises "the possibility of individual reverie . . . an oneiric [or dreamlike] pleasure".[9] In this light, it is interesting that Edensor (2010a, 199) develops his rhythmanalytic approach to understanding commuting by reference to certain arrhythmic experiences: "tyre burst, loud passenger announcements" and the *disrupted* daydream. Here are varied and indeterminate spaces that, in his analysis of traveling vulnerabilities, Bissell (2009b, 435) describes as producing "uncomfortable quiescence".

Such is the complexion of terror's reach into the armatures of commuting and its attack on dwelling-in-motion. Elden (2009, xxi) understands terror to be perpetrated by both state and nonstate actors who employ diverse practices—from the use of physical weapons to the deployment of "imagined geographies of threat and response". According to official definitions promulgated at s.2331, Chapter 113B of the US Code (United States Government nd), terror is specifically directed at noncombatants. It is intended to coerce, intimidate, and disrupt normal operations of government and populations—whether domestic or international, and draw international attention. Neither is it something assumed to be entirely 'foreign'. As Jeremy Packer (2006, 396) persuasively argues, turning "oneself into a literal bomb may produce ruptures" in control systems and societies with which one fundamentally disagrees and from which one seeks egress. However, any exit, escape, or removal from such systems and societies on the part of the suicide bomber is axiomatically self-destructive and ruinous of others. Moreover, and in the process, citizens "are made to fear the external threat of terror which legitimates their own treatment as a potential terrorist; as themselves a becoming bomb" (396).

The UK Government delineates the characteristics and reach of terrorism in a similar fashion: the Terrorism Act 2000 defines terrorism as involving or causing "serious violence against a person; serious damage to property; a threat to a person's life; a serious risk to the health and safety of the public; or serious interference with or disruption to an electronic system" (United Kingdom. Security Service MI5 nd). According to Anne Aly and Lelia Green (2010), such acts clearly instill fear; a debilitation pervasive, profound, and deeply experienced. These acts also engender high levels of psychological insecurity, which manifests as the search for safety and certainty, "ingroup favoritism, outgroup derogation . . . [and] increased rallying around the flag" (Orehek et al. 2010, 280). Discursively, this response is starkly illustrated by the conflation of several concepts: patriotism (loyalty to the 'fatherland' and one's fellow country-men); the sovereign (lord, ruler, master); sovereignty (preeminence); territory (in the sense of *territorium*, a place from which people are warned off) (OED); and, for example, by the full title of the USA PATRIOT Act 2001—which stands for Uniting and Strengthening America by Providing Appropriate Tools Required to Intercept and Obstruct Terrorism (United States Government 2001). In focusing on the relationship between terrorism and commuting as part of a larger analysis of the myopia of homeland security, James Mitchell (2003) outlines key characteristics of commuting in terms that are geographical, mobile, and rhythmic. There are, most obviously, daily shifts in large numbers of people that concentrate them in both motionless and moving sites and in situations highly susceptible to attack. Remember, as does Nick Paumgarten (2007, n.p.), that commuting

> is an exercise in repetition. The will to efficiency varies, but it expresses itself in the hardening of commuters' habits, as they seek to alleviate the dissipation of time and sanity. Some people travel with coffee; they have a place to buy it, a preferred approach to not spilling it, a manner of discarding the cup. You can spot the novice: he's rifling through pockets in search of his ticket, coffee bubbling up out the pinprick holes of his flattop lid, leading him to wonder how it is possible for the coffee to be leaking when the top is on tight. He has no strategy for newsprint stain. The pros have their routines. There's a group that plays bridge on the seven-fifty-eight to Grand Central. To get in a game during the short ride, they play speed bridge, a customized version with complicated rules. They often get into game-halting arguments about these rules, so they wind up playing less bridge than they would at normal speed. *Still, the fellowship, and the attempt at optimization, must bring some measure of happiness.* (emphasis added)

For individual commuters, the sense of fear and vulnerability in the face of terrorist attacks will depend on knowledge, the availability of protective resources, their access to alternative modes of movement, and the capacity of social institutions to enable choice about working and living arrangements. Moreover, "the cities that terrorists so often target are . . . homes, workplaces and economic engines . . . [and] highly specialized ecosystems, learning environments, symbolic territories, performance spaces and muses" (Mitchell 2003, 65)—in other words, they are places within which we *dwell even as we move*.

A later explanation on the legalities of subway searches by Charles Keeley III (2006) confirms that mass transit systems are key targets for terrorism, and considers the ways in which actions to prevent or preempt terrorist acts bypass the warrant and probable cause requirements in the Fourth Amendment (United States Government Ratified 1788). The work ends with Keeley (2006, 3233) proposing that the US Supreme Court adopt a *sui generis* exception to sanction anti-terrorism searches on subways on the proviso that search programs cede "no discretion to officials in selecting whom to search". In other words, the members of no particular group are to be targeted; at the same time, no one is to be trusted. There is no dwelling here.

The events of 9/11 have longstanding and complex genealogies (Elden 2009); equally, they signify originary moments in the war on terror and the modern state of terror. In work seeking to rethink terrorism and counter-terrorism in this knowledge, Bruce Hoffman (2002, 306) underscores the scale, lethality, and spectacular nature of the attacks. These were, he argues, "no prosaic and arguably conventional means of attack" and their lengthy planning, coordination, and execution underline the centrality of ancient ideas about martyrdom and modern methods of terror. Indeed, Hoffman refers to Osama Bin Laden as a sophisticated chief executive officer "grafting . . . modern organizational methods onto an ancient ideology of religious fundamentalism" (308) and, in doing so, invokes the words of Lawrence of Arabia, and his idea that men who dream during the day are dangerous because they will enact what they dream.

London's transport systems have been subject to multiple attacks since 1867. Yet, the bomb blasts on London Transport on 7/7 signify a shift in scale and approach. The reverie of commuters' dwelling-in-motion was shattered, the levels of endangerment and lethality of the explosions greatly increased by the contained environments of train and bus, and by rush hour conditions (Ryan and Montgomery 2005).[10] The events also underscore what Hoffman views as the escalation in psychological warfare following the attacks on the East Coast of the United States, and it is unsettling that he asks why terrorists had not thought to target mass transit more often. It is uncanny he does so; on 7/7 millions of people began their daily commuting patterns in and around London, unaware of a coordinated attack that was to unfold during the morning, the effects of which continue to reverberate in terms of securitization (Aly and Green 2010) and memorialization and commemoration (Allen and Bryan 2011). Four young men from Yorkshire and Luton—Hasib Hussein, Mohammad Sidique Khan, Germaine Lindsay, and Shehzad Tanweer—traveled from Luton to King's Cross, and then separated to detonate four organic peroxide devices which they carried in backpacks. Three bombs exploded within fifty seconds of each other at ten to nine. One, detonated by Tanweer, killed seven on an east-bound sub-surface train between Liverpool Street and Aldgate. One, killing six, was detonated by Khan on a west-bound sub-surface train just outside the bounds of the Edgware Road station as it was headed to Paddington. One, detonated by Lindsay, killed twenty-six on a south-bound deep-level tube train between King's Cross and Russell Square. The fourth device was activated by Hussein at nine forty-seven on the top deck of a London double-decker bus near the headquarters of the British Medical Association on Upper Woburn Place in Tavistock Square (British Broadcasting Corporation n.d.). The four explosions left fifty-six people (including the bombers) dead, and over seven hundred and seventy others injured; "the wider intimate reverberations are beyond quantification" (Allen and Bryan 2011, 263).

In work on the 7/7 attacks, Anna Reading (2011) considers the ways in which the mobile body bears witness to such events across multiple temporalities and spatialities, including commemorative ceremonies and activities held well after the fact, and she draws on Derrida to think out this idea of bearing witness. Derrida has himself been at pains to understand the act of witnessing, not in the least in *Sovereignties in Question: The Poetics of Paul Celan*.[11] In that work, Derrida (2005, 65) refers to one of Celan's poems, which recalls another kind of terror. In particular, he focuses on the following lines: "Aschenglorie hinter . . . Niemand zeugt für den Zeugen". In translation, these may mean *Ashes glory behind . . . no one testifies for the witnesses*. Derrida writes that ash is the name of what "annihilates or threatens to destroy even the possibility of bearing witness to annihilation. Ash is the figure of annihilation without remainder, without memory, or without a readable or decipherable archive. Perhaps that would lead us to think of this fearful thing . . . the capacity to bear witness" (68). He speculates that the term *ashenglorie* inaugurates a new body of truth: light from the fire of truth, but light that burns out, and "falls into ashes, as a fire goes out" (69). Of critical importance here, Derrida makes reference to the etymological relation between the term *witness* and the Greek *martyros*—from which derives martyr, the witness to cares and troubles, a word that, in Middle English, came to be the substitute for the Norse *pislavattr*: literally torture-witness (75; see OED). For Derrida, witnessing and proving are not the same: the former is based on faith and thus is always open to betrayal; the latter, *testimonium*, requires certitude. Returning, then, to Celan's idea that no one testifies for the witnesses, Derrida arrives at the point that neither the witness, nor witnesses to the witness, are ever actually present except through "what is called memory" (76). Brought into question by these ideas is the nature of what it is to believe; great terror and great trauma work to kill the capacity to dwell, move, and flourish.

THE RIGHT TO DWELL AND MOVE

Lefebvre (1996, 147) did not write his essay on the right to the city with modern forms of terrorism explicitly in mind; rather, his thinking was part of a larger agenda on space, time, and social life, on the "forms, functions and structures of the city". It outlines an emancipatory agenda he identifies at length.

> The *right to the city* is like a cry and a demand . . . [It] cannot be conceived of as a simple visiting right or as a return to traditional cities. It can only be formulated as a transformed and renewed *right to urban life* . . . place of encounter, priority of use value, inscription in space of a time promoted to the rank of a supreme resource . . . it gathers the interests . . . of the whole society and firstly all those who *inhabit* . . . [For who] can ignore that the Olympians of the new bourgeois aristocracy . . . are everywhere and nowhere. That is how they fascinate people immersed in everyday life . . . [Thus it is] essential to describe at length . . . the

conditions of youth, students and intellectuals, armies of workers with or without white collars, people from the provinces, the colonized and semi-colonized of all sorts . . . [It is] necessary to exhibit the derisory and untragic misery of the inhabitant . . . One only has to open one's eyes to understand the daily life of the one who runs from his dwelling to the station, near or far away, to the packed underground train, the office or the factory, to return the same way in the evening, and come home to recuperate enough to start again the next day. (158–9)

In place of this misery, the right to the city—indeed the right to dwell and access the benefits of urban life at its best—affords the opportunity for what Lefebvre called *Utopie expérimentale*. Here was a new urbanism based on a dialectical relationship between what is imagined and real; in it one senses the kinds of priorities that Wright (2013) has emphasized in work on real utopia, prosperity, and flourishing, noted in chapter one. Certainly, Lefebvre thought this new urbanism could produce "liberty, individualization . . . socialization, environs (*habitat*) and way of living (*habiter*) . . . leaving opportunities for rhythms and use of time that would permit full usage of moments and places, and demanding the mastery of the economic" (Kofman and Lebas 1996, 15, 19). The right to participate in civic life and the right to appropriate use values are intrinsic to Lefebvre's conceptualization; so, too, according to Kofman and Lebas is the need to understand difference—including by struggling against indifference and living differently. Lefebvre's ideas have been extended such that the right to the city is seen to embrace a series of conjoint claims. Mark Purcell (2013, 566–7) argues that these claims are "continually being remade as new groups initiate new claims in new contexts . . . against the background of those who have come before . . . In order to inhabit well—to realize a full and dignified urban life—their urban environment must provide them what they need". In this light, the stated motivations for the attacks of 7/7 both clash and jar. Here were young men who were, for all intents and purposes, struggling against what they viewed as criminal indifference. And yet, they perpetrated forms of violence on others that cannot—indeed, must not—be justified on any grounds. Thus, it is difficult to escape the irony of the message recorded by Mohammad Sidique Khan prior to the attacks, which were initially aired on Al-Jazeera television, and then reproduced by the British Broadcasting Corporation (2005a) in the September following the 7/7 attacks. Khan anticipated that the

predictable propaganda machine will naturally try to put a spin on things to suit the government and to scare the masses into conforming to their power and wealth-obsessed agendas . . . Our driving motivation doesn't come from tangible commodities that this world has to offer . . . Your democratically elected governments continuously perpetuate atrocities against my people all over the world. And your support of them makes you directly responsible, just as I am directly responsible for protecting and avenging my Muslim brothers and sisters. Until we feel security, you will be our targets. And until you stop the bombing, gassing, imprisonment and torture of my people we will not stop this fight. We are at war and I am a soldier. Now you too will taste the reality of this situation. (n.p.)

For me, this oratory brings to the fore the profound links between discipline and punishment as forms of conduct, and the geographies of misery and pain. It also underscores three key elements of the war on terror: prevention, deterrence, and preemption, on which Brian Massumi writes most powerfully. Massumi (2007) outlines the present distinction between prevention and preemption as follows: prevention has no ontology, having to assume that whatever is being prevented exists prior to intervention, and being a means to an end it cannot sustain itself and must "be leveraged from an outside source with outside force . . . borrowed power" (para. 5). Deterrence requires certitude because prevention has failed and threat is imminent, and it is this imminence that propels the apparent need for deterrence to produce its own cause to intervene. This cause, however, resides in the future and "to have any palpable effect it must somehow be able act on the present . . . as clear and present danger" (para. 8). Perversely, acting requires realizing rather than preventing the threat, and this becomes self-perpetuating. Thus, what "began as an epistemological condition (a certainty about what you and your opponent are capable of doing) dynamizes into an ontology or mode of being (a race for dear life) . . . Deterrence captures a future effect in order to make it the cause of its own movement" (para. 11). Preemption is also a present-tense, future-focused, operative logic, but no claims of certitude are provided: rather the characteristics of the threat are deemed incapable of being overcome: "threat has become proteiform and it tends to proliferate unpredictably" (para. 13). Therefore, Massumi concludes of this present logic, "You have to become, at least in part, what you hate" (para. 16); offence is seen to be the best defense. In such way, preemption gives license for unlimited conflict and, as Massumi poignantly remarks "the potential for peace is amended to become a perpetual state of undeclared war" (para. 25).

DROWNING NOT DRIFTING?

The Situationists sought to displace, dislocate, and see anew, using movement in ludic ways to rethink spatial and temporal relations. Their misappropriation and reappropriation represent a fascinating critique of modernity. So, too, does Brack's Melbourne painting; but while the multitudinous faces in his canvas may exude alienation and anomie, they appear neither distrustful nor afraid. In different ways, David Belle's journey home constitutes another playful critique of phenomena such as the rush hour, that signature point at which commuting's worst features intersect with poor urban design, the domination of transport infrastructure by motor vehicles, the use of time to regiment and then release workers onto streets and roads, and so on. Commuting clearly embraces variously mobile and rooted spatial and temporal relations, and creates or acquires all manner of things and events. It is riven through with diverse patterns, durations, instantiations, motifs, and regularities—rhythms that can be interrupted or disrupted. A consequence of these characteristics of commuting is that dwelling-in-motion is vulnerable to forms of conduct that would unravel the rootedness sought along the routes traveled each day on the way to work, however that might be conceived.

London's underground railways are sophisticated armatures created during the Progressive Era to secure massive profits to limited numbers of shareholders and fuel the engine of the economy. They came to be seen as phenomena that need not displace the so-called deserving poor and force the undeserving poor to move on, and that might enable widespread access to Pearson's (1859) incalculable blessing, suburbia. The first line to avoid the cut and cover method that was responsible for so much of the early dislocation experienced by the working classes of inner London, the Circle Line has been seen as a transformative part of the Underground. In turn, and in the hands of several charismatic owners and managers, the Underground itself has become a mode of movement elemental in the development of the passenger experience, that eurhythmic and spatially legible sense of dwelling-in-motion safely amongst trustworthy others: something, as Paumgarten (2007) suggested above, that might even bring some measure of happiness and enable spaces in which potential is not cut down.

Violence anywhere incorporates forms of disruption, misappropriation, and reappropriation that are contrary to flourishing even when the intention behind the violent act is emancipation. Anyone may be a potential bomb, we are told, and indeed so they may be. Some will become witnesses to acts of terror, either as victim-commuters who may or may not survive an attack, or as (what I will contingently call) victim-terrorists, young men in the main, *martyros*. Some will witness the effects of terror directly after the event as paramedics, loved ones, or bystanders. Of those, some may observe or experience the effects of suffering over weeks, months, and years. And all will witness, more generally, the erosion of trust: this, at least is evinced when my anxiety rises incrementally on any given day that I perform my Circle Line photographic essay under the gaze of surveillance cameras that I encounter on the London Underground. In my estimation, this loss of trust is immobilizing: it robs us all of the right to dwell and move, features of being human that are elemental in a flourishing life. In this respect, Massumi's prognosis, a perpetual state of undeclared war, remains intact and, in this instance, I can find no route home to notions of *eudaimonia*. How, then, to stay hopeful?

NOTES

1. It is challenging not to read Belle's disrobing as sexualized, given that he then executes his trace bare-chested, rolling several times on rooftop gravel and ignoring a bare-backed woman playing piano in one studio apartment whose garden he traverses. These devices simply underscore Belle's desirability as unattainable: he is moving at speed and in a disciplined state of 'flow'.
2. I argue that, no matter how novel, Belle's journey home may still be read as a form of commuting. Perversely, *Rush Hour* highlights how critically important to individuals, organizations, and institutions it is. Indeed, commuting's ubiquity does not, therefore, mask its complexity and entanglement in social, economic, and political life; the effect of public transport strikes is evidence enough of this.
3. Lefebvre (2004) is unlikely to have meant that rhythmanalysis is restricted to audition, and the importance of considering all the senses is made clear by, for example, David Bissell's work on vibration in train travel. Bissell (2010, 479) argues that vibration is an event with generative or productive capacity to give effect to new materialities, "uncertain and provisional connections between bodies, their travelling environments and the experience of movement where movement is not opposed to stillness".
4. Commuting (including its relationship to sense of place and ways of moving) certainly implicates children and young people (Stratford 2000a), a matter I raised in chapter one by reference to a conversation with a six-year-old student on the Piccadilly Line train to Heathrow. Nevertheless, my focus here is on the adult. How commuting is defined also affects understandings of volunteers, unemployed job seekers, retirees, or those involved in domestic/social reproduction who travel as part of their (unpaid) work, and has implications for trends such as telecommuting (Breaugh and Farabee 2012).
5. The German Socio-Economic Panel (G-SOEP) is "a longitudinal survey of approximately 11,000 private households in the Federal Republic of Germany from 1984 to 2011, and eastern German länder from 1990 to 2011. G-SOEP is produced by DIW Berlin. Variables include household composition, employment, occupations, earnings, health and satisfaction indicators" (German Institute for Economic Research (DIW Berlin Deutsches Institut für Wirtschaftsforschung e.V.) 2013, n.p.).
6. The British Household Panel Survey "began in 1991 and is a multi-purpose study [that] follows the same representative sample of individuals—the panel—over a period of years; it is household-based, interviewing every adult member of sampled households; it contains sufficient cases for meaningful analysis of certain groups such as the elderly or lone parent families. The wave 1 panel consists of some 5,500 households and 10,300 individuals drawn from 250 areas of Great Britain. Additional samples of 1,500 households in each of Scotland and Wales were added to the main sample in 1999, and in 2001 a sample of 2,000 households was added in Northern Ireland, making the panel suitable for UK-wide research" (University of Essex, Institute for Social and Economic Research 2013, n.p.).
7. Analysis by David Levinson (2008, 55) of panel data from thirty-three London boroughs over the decades between 1871 and 2001 also shows a "positive feedback effect between population density and network density" in suburbs, and "commercial development and concomitant depopulation" in central London.
8. Blake completed his own monotype of Newton in 1795 and it was not designed as a tribute; indeed, he was deeply critical of Newton's scientific materialism (Ault 1974).

9. It was that same year that Stanley also gained control of the London General Omnibus Company, an acquisition that created the Combine—integrated bus and rail travel—thereby seriously eroding direct competition. Noting that acquisition, Wolmar (2004) observes that London Transport marked two important changes in the story of rail: much higher levels of state control of the transport system of the national capital; and the advent of a public organization vested with both commercial and social responsibilities that, perhaps, would have satisfied the originary aspirations that Charles Pearson had shared in the London Tavern some fifty-three years earlier.

10. Ryan and Montgomery (2005, 544) explain as follows: small volumes of explosive translate almost instantaneously into large volumes of gas; an ensuing blast wave expands at the speed of sound causing four levels of injury. Primary injuries are typically at the interface of body and air, including the interior of the body (lungs, for example). Secondary injuries propel secondary matter into bodies; tertiary injuries propel bodies into solid matter; and quaternary injuries are caused by smoke and gas inhalation, heat and flame. "Confined spaces exacerbate such effects: surface reflections amplify and prolong the blast wave, the blast wind is channelled, and heat and gases are contained. The severity of injuries and the resultant mortality are thus greater".

11. Paul Celan was a Romanian German-speaking Jew and is among Europe's preeminent poets (Felstiner 2001).

6 Move It or Lose It

In chapter five, my focus was on examining how we move through cities and settlements—drifting, tracing, or dwelling-in-motion to the rhythmic rocking of railcars on train tracks—and on how forms of violent conduct, that are often (and qualifiedly) perpetrated by young adults, disrupt our rights to the places and spaces in which we work and live.[1] In this chapter, my gaze shifts again, settling on middle-aged and older adults, and my concern remains with questions of conduct and flourishing. Here, I want to examine and challenge the sorts of populist and scholarly narratives and practices that constitute aging as the inevitable, progressive, and sometimes rapid onset of decline and incompetence, and older people as increasingly marginal. I also want to consider various counternarratives of aging as positive and productive—a process in which flourishing is integral. In particular, this chapter is concerned with those aged from around forty to around seventy, although in a literal sense—and thinking back to the *0 to 100 Project*—the mid-age for Australian and Canadian men is approximately thirty-eight, and for Australian and Canadian women it is forty-one. However, life expectancy is increasing; indeed since "1970, men and women worldwide have gained slightly more than ten years of life expectancy overall, but they spend more years living with injury and illness" (Horton 2012, n.p.). Of course, although middle age and older adulthood are enumerated as an objective range, they are also qualitative categories[2] involving subjective—even secret—expressions of certain biological, psychological, and social rhythms (Lefebvre 2004). These subjective elements of aging are also captured in several narratives from the *0 to 100 Project*. Maria Jones, aged forty-nine, admonishes "Look after yourself. Eat healthy. Care about yourself. Keep yourself active. Keep positive. Smile". At fifty-one, William Buckingham observes that at a "certain point in my life, others will decide that my career is over simply because . . . my birthday is on somebody's HR database". Fifty-six-year-old Syd Haque suggests that what he knows now "with a 20-year-old body—I would be a lethal weapon", and Marianna Yeung, at sixty, simply notes "I can still walk, run and enjoy life, right?" In what follows, then, I ask *how is the aging body constituted, especially in terms of decline and increasing incompetence, and how have such notions been challenged, particularly by counternarratives of positive aging? How are certain geographies, mobilities, and rhythms implicated in regimes to produce fitness as an alternative to decline? What do these narratives reveal about what it means to flourish in middle and later life, and how and to what effects might 'moving it' constitute forms of conduct such as 'dressage' and training?*[3]

For fifty weeks each year and for thirteen years in a row, I have been watching what I eat, walking daily, and lifting weights four or five times a week. I introduced Pilates weekly into that routine about six years ago, and introduced transcendental meditation twice a day about four years ago. Over this period, it has been interesting to observe refinements to the sub-discipline of health geography that have augmented traditional models of population health with the consideration of well-being (Kearns and Collins 2010; Kearns and Moon 2002; Stratford 1998a, c). It has also been instructive to watch three trends. First, a growing number of resources overtly concerned with fitness have a strong subtext about flourishing and conduct—especially in terms of self-efficacy.[4] Consider the opening stratagem in the introduction to one instructional publication, *Strength Training for Seniors*, in which a focus on 'better' and 'improved' is followed by phrasing that seeks to unsettle the category 'old age':

> Combine [the statistics on aging in the US] with the undisputable science linking regular exercise with better health and improved quality of life, and you will understand why increasing numbers of people choose to stay actively well into so-called old age (or even to become *more* active than they were in middle age). (Fekete 2006, 1, original emphasis)

Second, many such 'how-to' publications—concerned with training, dressage, and discipline—share a concern to help readers understand risk, use information, become proficient, and reap the benefits of strength training and fitness programs. Thus, in another guide book, *Strength Training Past 50*, the benefits of investing in exercise regimens include increased sense of well-being in the world and an associated decreased risk of particular pathologies or morbidities (Westcott and Baechle 2007). Among those risks are decreases in metabolic rate, muscle mass, and bone density; increases in weight; the incidence of high blood pressure; cardiovascular disease; diabetes; musculoskeletal dysfunction; and gastrointestinal sluggishness.

Third has been the rise of the older adult role model of 'super' healthy and active aging. One notable example is Ernestine Shepherd, a septuagenarian from Baltimore, Maryland. Her story invites further consideration of Lefebvre's (2004, 39) ideas about dressage and rhythm, which he described in the following terms:

> Knowing how to live, knowing how to do something and just plain knowing do not coincide. Not that one can separate them. Not to forget that they go together. To enter into a society, group or nationality is . . .

also to bend oneself (to be bent) to its ways. Which means to say: dressage. Humans break themselves in [*se dressent*] like animals. They learn
to hold themselves. Dressage can go a long way . . . [and] bases itself on
repetition. (original emphasis)

On several sites that feature Shepherd's work, she relates an incident—being
in her fifties and shopping for swimsuits with her sister Mildred, who was
known as Velvet (see, in particular, AARP, 2011). The two had been commenting on their appearances, and Velvet decided that they should get active
and aim to be the oldest women bodybuilders listed in the *Guinness Book of
Records*. However, when Velvet died before reaching sixty, her younger sister
stopped training for an extended period, but later renewed her commitment
to the original pact. Shepherd's motto became "dedication, determination,
and discipline". She entered and won her first bodybuilding contest in 2008
and, by 2013, had secured several other competitions and awards, and been
featured in the *Guinness Book of Records* and *Ripley's Believe It or Not*.
Shepherd states that her goal is to let seniors "know it's never too late to be
fit in life". This was conveyed in a commentary crafted for a photographic
collection of inspiring over-fifties called "Beautiful Minds: Finding Your
Lifelong Potential" that was auspiced by the United States National Center
for Creative Aging and the DSM Nutritional Products' brand, *'life'sDHA'*
(Anonymous 2012, photo 10 of 21) (Figure 6.1). However, it is not simply

Figure 6.1 Ernestine Shepherd. Photograph courtesy of Beautiful Minds
(Beautiful-Minds.com). Reproduced with permission.

seniors she influences—one younger woman in her fitness class stated that Shepherd has shown her that "age is nothing but a number" (AARP 2011, at 1 min 39 seconds).

Shepherd's approach is self-evidently geographical, mobile, and rhythmic. For example, the spatial reach of her influence is made extensible by the Internet. Web sites and YouTube clips afford Shepherd the opportunity to disclose how her regime is informed by commitment to particular spiritual practices and ideas about "salvation" as well as her conduct as a role model—she explicitly understands her work as a form of ministry to help people to flourish (United Methodist TV 2012, 2 min 58 sec to 3 min 34 sec). Self-disclosure is important in that work, and Shepherd's daily routine, for example, is well-publicized in an interview in *Time Magazine's Health and Family* (Sifferlin 2013). Waking at the same time each day to dress in running gear, Shepherd will likely hear the patterned thud of her running shoe on pavement as she runs up to ten miles (sixteen kilometers) around streets and parks in Baltimore. Later, at the gym where she works out, she will count the length of time holding stretches; push and pull weights in a rhythm of concentric (muscle-shortening) and eccentric (muscle-lengthening) movements; circulate through a class to adjust others' posture or form; imbibe egg whites in a gulp; or meditate and ensure time to rest (see AARP 2011). In my reading, her routine and its publicity exemplify specific technologies of the self that demonstrate a clear understanding by Shepherd of the dynamics and appearance of her body; an appreciation of the influence of her workouts on her own sense of well-being and identity; and cognizance of the wider reach and effects of her mission to train herself and others.

The geographies of Shepherd's day are also gendered in novel ways. Although the free weights space in many mixed gyms has become less segregated over the last two decades, women still avoid what Lynda Johnston (1996) has described as the 'black and blue room', instead focusing on aerobic classes and other group exercise, or circuit training. Thus, in the gendered space of a 'typical' gym, Shepherd's use of heavy free weights also marks a boundary-crossing, and is additionally noteworthy because of her age. At the same time, in Shepherd's household there is a noteworthy reversal of heteronormative domestic roles (Gorman-Murray 2008), her husband intimating how he has transformed the kitchen of their home into something of a production line for the several highly prescriptive meals she eats daily (AARP 2011, at 4 min 10 sec).

To these mobilities and rhythms, others internal to the geography of Shepherd's body could be added: the flow of endorphins; the accretion of slow and fast twitch muscle fibers; growth in bone density; or shifts in respiration and circulation. Those micro-scalar 'events' in the body—those 'secret' rhythms—culminate elsewhere when Shepherd competes at national and international meetings. Her regime thus embraces the interior and exterior of the body, the domestic, local(e), and the international, and then folds on itself as she 'works' her embodied subjectivity. Such understandings point to two additional points underscored by Andrew Herod (2010, 76 and 78):

> i) the scale of the body itself only has functional coherence as a result of the coming together of myriad smaller- and larger-scale processes, for the body can only exist biologically as a result of what goes on within it and beyond it . . . and ii) the scale of the body as marked by its external features is itself not as discrete as is frequently portrayed in the social science and humanities literatures but is, instead, frequently crossed by chemical compounds, such that it is more difficult than perhaps first imagined to make the argument that the body is a distinct entity and that its covering of skin delineates where one space (the body's internal elements) ends and another (the outside world) begins. These facts raise questions of scale's gestalt . . .

I have experienced some such geographies, mobilities, and rhythms while working out in gyms in places as diverse as New York, Seattle, London, Suva, Portree on the Isle of Skye, Valletta in Malta, or Taipei. I have worked out in five-star hotels and grungy garages, and have a modest set of equipment at home—although I prefer the disciplined environment of the gym and the sense of camaraderie I feel with those working out around me. Although any set routine will only be observed for about four weeks at a time—the body becoming complacent if a workout regime is adhered to for too long—it has become habitual to move in specific ways at the gym and in Pilates classes. It has also become part of a narratological framework—a conversation carried on almost subliminally—to think like this: 'stand tall, hold the core, breathe deeply, be strong, eat this, drink that, do it now'. As David Bissell (2014b, 485) writes, "Habit understood as the development of . . . creative tendencies accounts for how movements over time and through repetition become transformed. What were initially movements that were clumsy, difficult and required effort, through habit became increasingly precise, graceful and effortless". I would not go so far as to say these characteristics typify my own routine; indeed, as I age, it becomes more challenging to maintain precision without lowering the weight of the barbell, dumbbell, or cable-machine. Nevertheless, the patterned and externalized rhythms of my days center on meditation and my workout; they are the first commitments in the diary and ones I am most reluctant to renegotiate. A recent annual medical checkup suggests that the internal rhythms of my body—blood pressure most especially—are 'gold standard' for my age group. At the same time, I have been deeply conscious of varied social, cultural, and spatial narratives that circulate in developed economies and underpin the creation of figures like Ernestine Shepherd, the promotion of the bodybuilding and the fitness industries more generally, or the notion that (interminable) youthfulness is *de rigueur*. I tend to read those narratives, and the discourses and practices that attend them, in terms afforded by Foucault's several works on the anatomo-politics of the body and the biopolitics of the population (see, in particular, Foucault, 1976). Thus, although my engagement with ideas about wellness, health, and fitness has been critical, still I am intrigued (and even drawn in) by the narratives I critique, the large and small changes in my body, and others' mixed reactions to my appearance, schedule, and habits.

All the while, I bear witness to my own aging. Some time ago, on hearing me grumble about the prospect of getting older, a dear friend some fifteen years my senior invited me to consider the alternative. Her dry humor brought home the paradox of what it means to age, the idea that aging is—in all likelihood—a privilege, and the reality that it remains a mixed blessing no matter one's setting. Over successive years, I have also seen many other older adults come and go in the gym, working out how to conduct themselves and construct new habits, rhythms, and skills, and forge new forms of etiquette and spatial relation, sometimes making long-term commitments, sometimes being present for a matter of days or weeks. In casual conversations at weight racks and water coolers, I have learned a lot about their motivation: almost universally, older adults attend to their physical fitness on the basis of a twin desire to stay well and healthy and have a 'good life'. Yet, the prevailing narratives of aging are those of decline and growing incompetence, which are attended by varied forms of discrimination, matters to which I now turn.

AGING AND THE DISCOURSES
OF DECLINE AND DISCRIMINATION

Implicating sites from the individual body to complex networks of social institutions across local, regional, national, and international scales, aging is a phenomenon critically important to attend to. In what follows, my concern is with middle-aged and older adults in developed economies. However, first it is useful to note the global reach of trend data on aging. According to the United Nations' Department of Economic and Social Affairs Population Division (2002), the aging of the global population is unprecedented, with increases in the proportion of those aged over sixty and decreases in the proportion of those aged under fifteen. The trend is largely attributable to the demographic transition, a longitudinal change that involves shifts from high to low fertility and mortality levels. In 1950, for instance, there were an estimated two hundred million older persons; around 2000 that had tripled to six hundred million; by 2050 it is predicted that there will be two billion older persons, another tripling. The fastest growth is among the oldest-old, who concern me in greater detail in the next chapter. There are also distinct gender differences in aging, with the Population Division reporting sixty-three million more women than men aged over sixty. Staff in the Population Division have also predicted that older persons will outnumber younger by 2050 for the first time, but they note that in advanced economies, that shift in relative proportions of older to younger people occurred in the late 1990s. Division staff have also identified several key implications of population aging, such as the rise in number of older persons and their proportional increase relative to those in the labor force. Among those repercussions are impacts on "economic growth, savings, investment and consumption, labour markets, pensions, taxation and intergenerational transfers . . . health and health care, family composition and living arrangements, housing and migration . . . voting patterns and representation" (United Nations' Department of Economic and Social Affairs Population Division 2002, xxviii).

Other data enumerate eight hundred and forty one million older persons worldwide, about two thirds of whom reside in developing economies (United Nations. Department of Economic and Social Affairs. Population Division 2013). Many live in poverty and have limited access to services, but where data are available, evidence suggests that significant numbers of "older persons make net financial contributions to younger family members until rather advanced ages" (xii). In advanced economies, most live independently until late in life, and in both basic forms of economy, developed (eight percent) and developing (thirty-one percent), there is a growing trend to continue working after sixty-five.

Notwithstanding such trends, aging is often understood as *necessarily* involving *irreversible* decline, and the creation of massive burdens on taxation systems and younger generations. Yet, longitudinal research on methods to arrest decline suggests highly promising results (Graves, Pollock, and Carroll 1994; Higgs et al. 2009; Kendrick, Nelson-Steen, and Scafidi 1994; Limacher 1994; Romero-Arenas, Martínez-Pascual, and Alcaraz 2013). Approaches have been identified that could increase incentives for individual and household savings and investments; ensure endogenous human capital formation—for example, by retraining older workers, and increase the age of retirement; and keep people engaged (Botman and Kumar 2008). Acknowledging the challenges involved in resourcing such methods, nevertheless, they are auspicious.

Arresting decline (and financing the means to do so) is often proposed as an alternative model of aging and as a process heralding the onset of greater levels of "stability in psychological wellbeing; continuing increases in vocabulary; greater selectivity in friendship and increased contact with close family; less need for novel stimuli; and increases in wealth, leisure time, and altruistic behaviors" (Albert, Im, and Raveis 2002, 1214). Even so, such prospects are seen to necessitate significant public health and government interventions to delay or prevent the onset of frailty and isolation, for instance as a result of the cessation of walking or driving (Satariano et al. 2012). Under consideration are individual and collective lifestyle choices, living arrangements, and their effects; diverse risk factors; and the capacity to function by modifying social, psychological, and physical environments, and by creating or adjusting technologies. For example, in research evaluating the efficacy and efficiency of London's transport networks for older adults, Laura Ferrari and her colleagues (2014 in press, n.p.) have suggested that the "intersection of limits to government support with the growing mobility needs of the elderly and of people with disabilities calls for the development of tools" that allow policymakers to decide how to prioritize investment. Acknowledging that buses and trains have been reengineered for increased accessibility, they point to the significant costs yet to be borne as a result of modifying interchanges and stations, the costs of which are predicted to rise. Similar observations hold in relation to the built environment (Antoninetti and Garrett 2012).

Indeed, on the basis of global and national models of present and predicted burdens of inactivity and ill-health, as well as burgeoning services for health care, aging will be a profound fiscal challenge (Beauchamp et al. 2007; Dishman 1994; Gordon and Cuttic 1994; Graves, Pollock, and Carroll 1994). Nevertheless, it is recognized that substantial consumer choice is likely to be available to middle-aged and older adults in middle and upper income brackets, or in societies with amenable structural systems—including in relation to health and well-being services. Simon Biggs (1997, 555) argues that, when able, people will use that power as part of a mask or 'masquerade'—a set of practices to recode the body that emphasizes "the fluidity of identity choice". Yet, on the one hand, masquerades in defiance of *appearing* older are not necessarily equivalent to other technologies of the self that work to enhance fitness and well-being, and their effects may literally be skin-deep.

At the same time, Owen Jones (2012, 3) notes that one London fitness club offering classes in 'Gymbox' required "a steep £175 [US$285–90] joining fee on top of £72 a month for membership" and suggests that the program was "launched to tap into the insecurities of [the gym's] predominantly white-collar professional clientele". The example of the London club is in marked contrast to the operation of working-class gyms (Heiskanen 2012), the geographies of which are often defined as masculine spaces oriented to boxing and heavy weight lifting, and which are sometimes highly differentiated according to ethnicity. Nevertheless, as Lucia Trimbur (2011) has established, such spaces can effect shifts in overt and tacit discourses of oppression and disadvantage, especially under the guidance of strong trainers. In Trimbur's research, these trainers express confidence that their clients are able to produce new and different forms of identity, and they "reject disadvantage as the defining feature of postindustrial subjectivity as well as the collective assignment by neoliberal politicians and policymakers of men of color to a category constituted by a pathological cast of characters—damaged fatherless men, deadbeat dads, deviants, hustlers, gangsters" (350).

The experience of exercise at the London fitness club also differs significantly from the ways in which working-class women, for instance, use their bodies to labor, simultaneously mobilizing particular understandings of the relationship between the corporeal, class, and capitalism. For example,

Alex Dumas and Suzanne Laberge (2005) have established that many working-class Canadian women in their study are little inclined to exercise on the basis that they labor with their bodies. In this respect, Andrew Herod (2010) refers to four apt understandings of this laboring body that are indebted to Marx, and that echo both Susan Marston's (2000) concerns with scales of production, reproduction and consumption, and Richie Howitt's (2002) consideration of scales of dispossession, to which I referred in chapter one. Herod (2010, 69–73) views

> the human body as the ultimate source of all transformation of the material world and thus as the original source of wealth . . . [as] dramatically impacted by the conditions within which it lives, such that metabolism and biology can be significantly shaped by the social relations of work . . . [as individual] workers . . . rendered increasingly invisible as the capitalist system had developed historically . . . [and as] a model for understanding how capitalist economies work" in terms of circulation.

Thus, identity choice is constrained geopolitically and often markedly among women. For example, those women in the United States aged fifty-five to sixty-four years of age who depend on spousal health insurance coverage often experience a "window of vulnerability" in terms of health care and services if their partners retire or die (Angel, Montez, and Angel 2011, 6). More generally, there is a strong, positive correlation between income and class, and the experience of locational advantage or disadvantage, including in relation to housing, neighborhoods, environmental pollution, crowding, and noise exposure (Evans and Kim 2010). These correlates exacerbate multiple risk factors at all cohorts and are confounded by aging.

Notwithstanding, body image is seen as critically important over the life-course (Baker and Gringart 2009), and the body is constituted as a physical and moral project (Andrews et al. 2012; Rudman 2006). Justine Coupland (2009) has illustrated the reach of consumer practices in relation to 'anti-aging' in an analysis of narratives circulating in *Woman and Home* and *Good Housekeeping* magazines from the mid-2000s. Coupland notes, for example, the ways in which time is constituted as etching the contours of one's facial terrain—it is a rhythm with physical (indeed spatial) tracery that can be conquered or masked by using particular 'age-defying' skin products, for instance. Time also engraves certain expectations on one's conduct, often in decadal rhythms (Nikander 2009). It is commonplace for both men's and women's magazines to now feature articles about hair, makeup, beauty regimes, diet, or exercise on the basis of a passage through the twenties, thirties, forties, and fifties. Thereafter, and in relation to the sixties and later decades, silence often ensues—as if to infer a human equivalent of the 'best by' or 'use by' date stamped on consumer goods. This silence is instructive, and highlights the existence of an oppressive and double-edged process. On the one hand, strong social and cultural messages compel people to engage in individualized quests to be 'not old', and these may or may not be allied to other messages about health and a sense of flourishing (see, for example, Flourish Over 50 2014). On the other hand, there is a widespread tendency to ageism (Calasanti 2005).

My contention remains that ageism is directed against not only those who are older; were it so, there would be little merit in the argument that young children are more than 'human becomings', as posed in chapter three, and there would be limited worth in the idea that generosity towards teenagers and young adults could characterize urban design, as suggested in chapter four. Rather, ageism is spatially and temporally contextualized, and involves many forms of discrimination that are expressed as attitudes and actions that diminish, disadvantage, or exclude individuals on the basis of their age. Moreover, ageism may manifest over time as a complex *internal* wrangling with one's 'moral compass' in diverse contexts. For example, a skater might, in midlife, find herself an urban planner exercising professional authority by making key policy decisions that perpetuate local design systems the geographical effect of which is to marginalize skating. Her decision might draw on analysis showing a preponderance of residents over sixty in the locale/s in question, community concerns about their mobility, and the existence of shortfalls in public funds that prevent innovative design systems which might accommodate varied forms of rhythmic urban play. Later, in retirement, she may find herself speaking out at public meetings to counter claims that all skaters are disruptive and dangerous, or that parkour classes in local parks designed for the elderly are a waste of ratepayers' funds, and might then be in a position to lobby for more creative urban design outcomes and participate in community grant applications to advance such ends. In more general terms, "chronological age and lifespan categories and other inter-actional formulations of age surface and are made relevant for and by us, implicitly or explicitly, as we position each other or describe and account for our own and others' actions in various everyday settings" (Nikander 2009, 864).

'LIFE BEGINS AT' . . .

In contrast to narratives that constitute aging as necessitating decline and diminished competence are a growing number of other approaches variously labeled 'positive aging', 'aging well', 'productive aging', and so on. In material on the Australian Psychological Society's website, for example, it is evident that positive aging is seen to embrace diverse values, beliefs, principles, attitudes, behaviors, and practices. It is assumed that older adults can and will take responsibility for "feeling good . . . keeping fit and healthy, and engage fully in life" (Australian Psychological Society 2014, n.p.). In the face of significant change and challenge, positive aging is positioned by the Australian Psychological Society as enabling "a sense of control" in new parts of the life cycle, and as providing certain dividends because those "who age positively live longer and healthier lives, and enjoy a good quality of life" (n.p.). Particular indicators of positive aging are noted: maintaining an optimistic attitude, staying connected, keeping the brain active, managing stress, engaging in volunteer or part-time work, engaging in physical activity, having regular health checks, and eating a healthy diet. The Government of Canada National Seniors Council (2011) provides an almost identical summation of positive responses to aging, as does the Positive Aging Resource Center, funded by the US Department of Health and Human Services (Brigham and Women's Hospital Boston 2014). Similar information is promulgated by at least one set of UK providers of positive aging courses, the inference being that personal development techniques—learning how to conduct oneself positively through aging processes—"can help us all develop more resilience and strengthen our sense of choice and control" (Positive Ageing n.d., n.p.). Acknowledging normative narratives of aging-as-decline, Susan Paulson (2005) asserts that mobility loss signifies dependency and helplessness, and represents the aging body's low point. Her own work demonstrates how fitness and dance class participants seek to reinscribe their approach to growing older by doing so in ways that are healthy and engender a sense of well-being, and shows how they derive different benefits such as individual physical fitness and collective social connection.

One standout paper on aging, aging well, and mobility by Susan Nordbakke and Tim Schwanen (2013, 1) brings together a review and analysis of the literature on these concerns to demonstrate longstanding interrelationships among them. Their work is salient for my purposes because its central focus is the idea of well-being as a form of flourishing based on a commitment to purpose. Nordbakke and Schwanen describe mobility as both embodied movement and motility and the potential for movement on the basis that they implicate physical and psychological functional capacities that underpin "the good life" (2). Although their focus is on older adults above retirement age, the arguments they advance are more widely applicable. For example, their thinking about well-being is developed as an heuristic framework along three dimensions: is well-being understood subjectively or objectively, as hedonic or eudaimonic, or as universal or contextual? In relation to the first of these, people seem able to perceive and experience a sense of subjective well-being even if various objective measures indicate that their circumstances suggest otherwise. In terms of the second, they "calculate their utilities [measures arising from a capacity to satisfy preferences] and try to maximize their rewards in terms of happiness (as opposed to pain and suffering)" (4); such hedonic tendencies are in contrast to the idea of *eudaimonia*. In relation to the third, the idea and experience of flourishing is understood by universalists as being an essential property with a limited set of characteristics, or as possessing "a minimum of common conditions that are valuable to all humans, independent of time and place" (5). Alternatively, contextualists argue that well-being is necessarily influenced by prevailing conditions, and is simultaneously a social process, a technology, and a discourse.

Nordbakke and Schwanen then apply their understandings of these debates in a number of ways[5] and specifically in relation to movement and motility. They suggest that among older people, there are strong correlations between the capacity to be mobile and the experience of well-being; they argue that the different approaches are complementary and have greatest explanatory power when considered together. Three recommendations about studies focused on aging, well-being, and mobility are then advanced by them. First, subjective studies are critically important, but should be augmented with objective investigations which are "helpful for identifying on the basis of explicitly normative criteria groups of older adults for whom interventions through public policy is warranted" (21). Second, hedonic studies need to be amplified and enriched by others that emphasize *eudaimonia* on the understanding that they "afford a richer set of hypotheses about how potential and actual movement can contribute to well-being in later life" (21). Third, because whether and "how flourishing ensues depends on individuals' abilities and the social contexts in which they are embedded" (21), more contextual studies are recommended. In particular, ecological and geographical studies are considered useful in helping researchers to ground the insights they derive from studies of well-being, aging, and mobility.

Older adults committed to fitness practices exemplify the power of a composite approach such as that which Nordbakke and Schwanen envisage. As an expression of well-being, fitness is ontology, discourse, dressage, and geography. It is, for example, concerned *both* with feelings, experiences, and perceptions *and* with how one measures up—in quantitative terms understood as sets, repetitions, aerobic fitness, strength, or endurance. It is concerned with training in ways that maximize benefits such as fitness and that minimize disadvantages such as pain; in the terms used by Nordbakke and Schwanen, it is hedonic. Equally, fitness is strongly oriented to complexly configured purposes beyond mere utility and is thus eudaimonic. Ernestine Shepherd, for instance, discloses that her sense of well-being derives partly from being fit, strong, agile, and aerobically fit; partly from being seen as attractive and desirable; and partly from being an international role model and celebrity.

However, a sense that fitness and flourishing work in tandem is not simply restricted to the geography of the body; many of those who train state they are attached to their home or community gym, or their running or walking route, in ways that gesture to sense of place, and to dwelling-in-motion. In this light, it is noteworthy that Karolina Doughty's (2011, 148) work on walking's rhythms includes an examination of the relationships between healthy bodies, place, and dressage, and embraces an argument that the "concept of rhythm inspires a perspective on the embodiment of the new public health as an assemblage of intersecting rhythms". Indeed, Doughty posits that "the promotion of a particular form of 'health walking' informs individuals of 'appropriate' modes of walking based on embodied norms and repertoires of movement and their physiological responses" (148).

The relationship of fit aging to flourishing gains expression in other ways, too. For instance, Toni Calasanti (2005) argues that gerontologists and those in allied and other fields have tended to constitute positive aging as a process and outcome of healthiness, self-control, and moral goodness; this stance seems to exist whether aging is viewed as purely biological, socially constructed, or multifaceted. In elements of his analysis of biomedicine, power, and subjectivity, Rose (2007) underscores some of the effects of such ways of thinking. His project, to which I referred in chapter two, seeks to interrogate widespread appeals to "a transcendental religious morality or an equally transcendental human ontology" (2). In particular, Rose points to a step change in the technologies of the self that implicates varied scales from the molecular to the societal, and that enables "a qualitative increase in our capacities to engineer our vitality, our development, our metabolism, our organs, and our brains" (4). By such means, Rose points to the existence of ideas of genetic personhood "that construct the subject as autonomous, prudent, responsible, and self-actualizing" (129). Arguably, adults in middle and older life who seek to 'move it or lose it' are among those so constituted. For my purposes, it is apt that Rose writes of the doctrine of soteriology as "a way of making sense of one's suffering, of finding the reasons for it, and thinking of the means by which one might be delivered from it" (255). Rose also refers to various historically and geographically contingent forms of salvation that, he suggests, have guided our conduct as we attempt to overcome our general unease. Much such ethical work, Rose maintains, has been done in the name of "vital existence" (255), increasingly in ways connected to work in biotechnology and biocapital. Yet, recall Ernestine Shepherd's assertion that her work is spiritually motivated and informed by specific ideas of salvation. In such terms, one could view older adults in the gym as seeking deliverance from the suffering of age and of ageism, and as having accepted both explicit and tacit scripts about being responsible for their health. In these labors, bodies serve as key morally-charged signifiers of larger investments in technologies of the self—intended to prolong middle age and foster opportunities to flourish (Wright, C. 2013).

Cassandra Phoenix and Bevan Grant (2009) note that many studies of physical activity are centered on the body, and on corporeal and experiential responses to various regimes. They emphasize the worth of telling stories on the basis that "while making sense of our actions and experiences, we not only tell stories about our aging bodies but also tell stories out of and through our aging bodies. Consequently, the body is simultaneously cause, topic, and instrument of whatever story is told" (369).[6] At the same time, they concede that many investigations focused on the body are useful for understanding balance, strength, functional capacity, recovery, aerobic fitness, bone density, cholesterol, types and amounts of exercise needed, and so on—some of which I return to shortly. Nevertheless, they make the point that the older body is far from being a straightforward stimulus-response mechanism; rather, it is ambiguous and its challenges are intricate (which, of course, is true across the life-course). On this basis, they argue that it is insufficient to consider the body alone, a practice that reinscribes the apparent primacy of prudential individualism and consumerism, and does not account for structures, policies, and personal and societal conditions.

Phoenix and her colleague Brett Smith (2011) later suggest that some of the practices of older, natural[7] bodybuilders indicate strong resistance to a 'sentence' of senescence, dominant narratives of aging as decline, and the medicalization of aging bodies. They draw on narrative and dialogical analyses of life-story interviews conducted by them over three long sessions with thirteen such bodybuilders between fifty and seventy-three years of age. Of those, two are women, a number approximating the proportion of those involved in natural bodybuilding in wider populations. In their work, Phoenix and Smith (2011) pinpoint among participants a universal tendency to tell what they call counterstories to aging as decline, including the onset of disease and dysfunction, or the tendency to disengage.[8] Those counterstories suggest that the bodybuilders are engaged in creating narratives of self-empowerment, awareness, and care (see also Baker and Gringart 2009; Eman 2012; Higgs et al. 2009; Phoenix 2010; Rudman 2006; Umstattd and Hallam 2007). For example, Phoenix and Smith report how participants would speak of the effort to be motivated and "keep plugging on" (633). Such discipline requires living close to or within reach of a gym, or committing to travel to one, as well as organizing one's social, temporal, and spatial life around working out. It requires being willing to act as a role model—Phoenix and Smith referring to one sixty-three-year-old bodybuilder who related stories of being used by staff as "a specimen" so that other gym clients "can't use the usual excuses about being too old to do it [work out]" (633). It further necessitates a capacity to resist negative stereotypes and challenge prevailing ideas of aging-as-decline by being *visible* in different spatial contexts; that is, by being mobilized. As one participant described it to the researchers, this work is "not just [about] telling but also showing a counterstory by overtly occupying spaces such as the dance floor, a school, the free weights area of the gym [which is dominated by young men who often are very strong and bulky], and . . . on the natural bodybuilding competitive stage" (634). For Phoenix and Smith, such stories are intrinsically interesting. More importantly, perhaps, they are instrumentally significant: forming part of a *narrative habitus* fashioning how we view and understand our lives, conduct ourselves, and flourish, the accounts have the capacity to influence policies and programs on aging.

Returning, finally, to Phoenix and Grant (2009, 364), it is noteworthy that they demonstrate how discussions a decade ago about the aging body were merely emergent in social theory, and suggest there is room to augment the discussions that have ensued in the interim because "every body's story needs to be considered and developed if we are to realize how various aspects of aging and physical activity . . . can be best represented". The insights that they share point to opportunities to better comprehend the relationships between aging, well-being, and mobility, especially given medical and allied research suggesting that *life begins at* the age people start to invest in dynamic forms of self-care, even if that transpires in later decades. I want, now, to consider some of the effects of such investments.

GET WITH THE PROGRAM . . .

In early work on the geographies of fitness, Derek McCormack (1999, 165) noted that they are "woven from the shifting landscapes of risk, scientific knowledge and flexibility" that implicate varied spaces—laboratories, marketing firms, fitness clubs, and homes—as well as the intertwined sites of bodies and machines. His work in this respect was, I think, prescient in relation to fitness and the aging body.

   Extended longevity is a portmanteau idea for a range of key uncertainties in governing—the conduct of conduct. Insights from Patrick O'Malley's work on risk continue to be instructive in this regard. O'Malley (2000, 461) has proposed that uncertainty is a modality of governing that "relies both on a creative constitution of the future with respect to positive and enterprising dispositions of risk taking and on a corresponding stance of reasonable foresight or everyday prudence (distinct from both statistical and expert-based calculation) with respect to potential harms". In short, uncertainty is intended to spur action in the form of calculated risks—starting an exercise regime, changing one's diet, or recalibrating one's identity in line with changed body image. As Rudman (2006, 190) suggests, practices of body optimization

> generally involve fitness activities and body modification techniques. Articles [in this instance in Canada's *Globe and Mail*] variously present consumer-based activities and surgical options, such as cosmetic surgery, rock climbing, and exercise clubs that 'add years to our life and add life to our years' . . . and 'turn back your biological clock' . . . The 'prudential consumer' is depicted as a person who, beginning in mid-life, proactively maximises his or her sense of security for the future.

More broadly examining uncertainty in neoliberal government, O'Malley (2000, 465) also notes how the alignment of uncertainty and risk has, since the nineteenth century, constituted a "hybrid of enterprising prudentialism" when statistical calculation emerged as a science of governing. O'Malley speculates that prudent subjects are meant to remain autonomous by, for example, assembling information, materials, and practices into personalized strategies that identify and minimize their exposure to harm: note how different in tone and intention this idea is from optimizing one's exposure to conditions that engender flourishing. But O'Malley (2000) goes further, by considering whether there is no liberty to create the future; rather, there is only the freedom to choose from an already calculated set of products that engender a responsible lifestyle. Notwithstanding, prudent subjects may also be risk takers "imagined as innovators, who 'reinvent' themselves and their environment. Here they appear as entrepreneurs, not as prudent consumers of risk" (465).

Although there is mounting evidence of the efficacy of particular ways of being and doing in relation to health and fitness, to conduct oneself in order to invent the future in ways that are not statistically calculable is to remain uncertain (and one can imagine Aristotle arguing that this uncertainty is insufficient reason to cease striving). Yet, as part of self-care, extensive benefits are seen to flow from moving. A first—although highly qualified—advantage appears to be a pronounced capacity among older adults who exercise to self-regulate, manifest self-efficacy, and express certain 'outcome-expectancy values'. As Umstattd and Hallam (2007, 207) describe these three ideas, self-efficacy "is perceived confidence in one's ability to perform a specific behavior in a given setting . . . Self-regulation is the personal regulation of goal-directed behavior or performance . . . Outcome-expectancy value is the interaction between a person's estimate that engaging in activity will lead to a certain outcome (outcome expectations) and how much the person values that outcome".[9] Their research suggests that, of the three, self-regulation "might be a more influential construct in predicting change in exercise" (214). Of note, it comprises several forms of conduct, including "self-monitoring, goal setting, social support, reinforcement, time management, and relapse prevention" (209)—all of which Ernestine Shepherd appears to exemplify in spades, for example.

Second, there is evidence for improved functional fitness on the commencement of exercise, even amongst those aged over sixty-five classified as being at high risk for the loss of muscle strength (Fahlman et al. 2007). Exercise regimes comprising three to five sessions weekly, and involving combinations of warm-up, aerobic work, resistance and/or circuit weight training, and cool-down appear to provide significant functional improvements over a matter of weeks (Graves, Pollock, and Carroll 1994; Romero-Arenas, Martínez-Pascual, and Alcaraz 2013). The adoption of such regimes will often increase levels of endorphins in the blood, as well as endurance and muscle mass, and enhance oxygen uptake and a generalized sense of well-being. For my purposes, such findings are noteworthy because the patterned rhythms of functional movement are shown to be immensely helpful for balance and proprioception, which are critically important in mobility. Derived from the Latin *proprius*, meaning one's own perception, the latter term provides crucial sensory information about the position of various body parts and the effort that is being employed in movement and acceleration (OED). Middle-aged and older adults tend to experience a diminishing sense of balance, a key signifier of reduced proprioceptive capacity that can lead to lower levels of confidence about mobility and motility, growing concerns about safety, and a tendency to reduce simple movements such as walking, which are clearly linked to sound health. Proprioception is also negatively affected by (a) neurological conditions—which affect the capacity to shift the weight from one foot to another, (b) diabetes—which damages the nerves in the foot, (c) vertigo brought on by Ménière's disease—often precipitated by excess salt, alcohol, or caffeine consumption, (d) postural hypotension—which causes light-headedness on standing, (e) other foot problems—corns, bunions, and hammertoes, for instance—which are exacerbated with age, and (f) certain medications—not least are those involved in treating hypertension or high blood pressure.

Many of these issues are experienced as mature-onset conditions. Nevertheless, and using the adage 'use it or lose it', a report from the Harvard Medical School (2006) recommends particular forms of conduct to retain or regain balance: core strength and good posture, and exercises that improve equilibrium such as tai chi or the use of rocker boards, balance balls, or rotation disks. The report also points out what is at stake here in terms of flourishing: "Every year, more than a third of people over 65—and half of those over 75—take a tumble. Falls account for about 300,000 hip fractures annually. For older people, they're the leading cause of death from injury and a major cause of disability" (1). Here, it is important to acknowledge not only that people experience the onset of disability with increasing age, but grow old in bodies that have been deemed disabled, including from birth. Gavin Andrews and his colleagues (2012, 1928) draw attention to the ways in which, for people with disabilities, movement is "an intensely embodied and emotional experience, involving bodily strain and effort, tiredness and the need to rest, and self-consciousness of moving (perhaps differently) through public space". But on the basis that movement "can boost self-confidence and esteem, and build a sense of belonging and right to being in public spaces" (1928–9) they also stress the importance of claiming the right to do so.

A third dividend from an investment in self-care is pinpointed by Gary Gordon and Marianne Cuttic (1994) in work on the aging foot, which might also have fascinated Lefebvre, de Certeau and Ingold. Without care, the aging foot is subject to decreased joint mobility, fat pad atrophy and associated degeneration of collagen, increased bone stress, and impaired circulation. As well as compromised balance, such symptoms can be critically disabling of 'normal' mobility, as recorded in tests to measure different modes of standing, moving from sitting to standing, and gait patterns (Sargent-Cox, Anstey, and Luszcz 2012). Where competency in these modes is reduced, one's daily rhythms and geographical range may be truncated; conversely, where they are maintained instances, durations, and patterns of movement are likely to be enabled, and range preserved. Pilates is one set of movements whose instructors emphasize the importance of the foot for well-being and for which there is evidence of improvements in balance over time (Bird, Hill, and Fell 2012).

Better muscle mass and psychomotor performance appear to be a fourth composite benefit of self-care (Graves, Pollock, and Carroll 1994), and evidence suggests it can be regained after long periods of relative inactivity. Lean muscle mass is a term describing the weight of muscle in a body; its presence is associated with healthy connective tissue and bone density crucial for reducing the risk of injury from moving. Appropriate intake of protein, estimated at or near one gram per kilogram of body weight, seems elemental in maintaining muscle mass. It has been thought that the more muscle one has relative to body fat, the higher one's basal or resting metabolic rate—or BMR (Kendrick, Nelson-Steen, and Scafidi 1994), and there is a prevailing idea that muscle mass and associated strength decrease naturally at about five percent per annum from about the age of thirty-five. This process has been thought to result in metabolic slowdown, which may lead to weight gain and negative effects on organ, skeletal, circulatory, and other system functions, and may affect strength and performance. Alternatives exist; not least among them is maintaining muscle, and that requires exercise—and most typically resistance training of some description (Fahlman et al. 2007; Romero-Arenas, Martínez-Pascual, and Alcaraz 2013). When crafted as compound movements rather than isolated ones, the effects appear more pronounced. For example, one might do a triceps kick-back and affect the triceps muscles at the back of the arm. Or one might do a bench press and bring into play the chest or pectoral muscles, as well as triceps, deltoids or shoulder muscles, and the biceps at the front of the upper arm as stabilizers. These compound movements appear to improve balance, optimize blood sugar levels and sleep patterns and, again, give a generalized sense of well-being. In my experience, they also require individuals to understand their bodies as one might a well-loved landscape in which one dwells, and to know how that landscape works, how things move within it, how diurnal and other rhythms affect it, and how it changes over longer periods of time.

A fifth value of self-care derived from observing the principle of 'move it or lose it' is better cardiovascular fitness (Agarwal 2012; Limacher 1994), which is commonly understood as the capacity of the heart and lungs to supply oxygenated blood to working muscle tissue, which then uses the oxygen, among other things, to produce energy to move. Marian Limacher (1994) points to two influences of aging on cardiac structure. First are microscopic increases in fat, collagen, lipofuscin—a yellowish granular pigment that appears to be a by-product of oxidation of unsaturated fatty acids, amyloids—misfolded and fibrous proteins that disrupt the architecture of deep tissues, and the size of myocytes—long tubular cells with different functions found in different kinds of muscle tissue, including the beating heart—which also decrease in number. Second are anatomic changes: the elongation, and an increase in the diameter, of the heart's main artery, the aorta, the wall of which also thickens; increases in the dimensions of the left atrium—the upper left chamber; an increase in the left ventricular wall and concomitant decrease in the size of its cavity diameter; and the calcification of mitral and aortal valves and their annular rings, which may affect directional blood flow. Yet Limacher (1994, S15) provides evidence that, in terms of maximal oxygen consumption, "older athletes who maintain competitive levels of training show only half the expected rate of decline" and confirms that "integrating exercise into one's lifestyle, regardless of the age at which one begins, will also promote overall health benefits". According to Candice Hogan, Jutta Mata, and Laura Carstensen (2013), such benefits extend to cognitive performance—including better performance on working memory tasks, and high-arousal positive affect—a heightened sense of anticipation and reward. They also include improvements in bone density, without which mobility is at risk from fractures (Eastell 2013).

A sixth dividend of fitness as a form of self-care appears to be a reduced incidence of disease (Lowenthal et al. 1994). One reason for this correlation is that older adults engaging in exercise tend to avoid the onset of obesity, which may aggravate "age-related decline in insulin sensitivity and [be] associated with risk for cardiometabolic syndrome"[10] (Bouchonville et al. 2013, n.p.). Synthesizing a number of studies that involve interventions, longitudinal timeframes, and cross-sectional approaches to sampling, Louis Bherer, Kirk Erickson, and Teresa Liu-Ambrose (2013) have established that physical exercise also positively affects cognition and tends to delay or halt both its decline and the onset of neurodegenerative disorders such as Parkinson's disease.

There is, then, longitudinally significant evidence that supports the correlation between fitness and movement. That evidence also throws up varied problematizations implicating temporal rhythms such as spans of decades, different ways of moving, and numerous geographies both internal and external to the body and which, on closer scrutiny, suggest more complex assemblages at work. Entangled in the ideas and practices that underpin the phrase 'move it or lose it' are other questions about what it means to conduct oneself, to train, to constitute, reconstitute, define, and simultaneously resist the contours of interiority exteriority, of body self-identity, to serve as a model for others and, on occasion, to couch all such acts in terms of a doctrine of salvation. This last compulsion stems, in no small measure, from the inevitability of having to bear witness to our own demise and it is to that matter that I now turn.

NOTES

1. Profoundly uncomfortable and contentious although it is to say so, these observations are not intended to deny the fact that some of those who are propelled into lives of violence feel that such actions are the only means in which to seek ways out of situations they deem ethically, morally, physically, or in other ways insufferable.

2. For example, the British Broadcasting Corporation (2012) has published a report referring to a thousand-respondent 'Love to Learn' survey, findings from which suggest many Britons view middle age as beginning around sixty years of age *or* when one's demeanor shifts from a youthful approach.

3. Lefebvre's writing on dressage, I think, parallels ideas that Foucault promulgated about the conduct of conduct and technologies of the self. Let me explain by reference to Elden and Crampton (2007, 7) and their identification of Foucault's understanding of government as shifting from "a simple retention of territorial control to a more nuanced notion of government over a 'complex' of men and things constituted as a population . . . an object itself, with birth and death rates, healthiness and so on". Such matters concern me through the chapter, and have been built upon by scholars such as Rose (1998, 2007) and O'Malley (2000), to whom I also refer here.

4. In the *Ethics*, Aristotle (350 BCE(b)) makes numerous observations about exercise and health. In Book II:1, for example, he notes that "both excessive and defective exercise destroys the strength, and similarly drink or food which is above or below a certain amount destroys the health, while that which is proportionate both produces and increases and preserves it. So too is it, then, in the case of temperance and courage and the other virtues". Exercising care with one's health, being moderate in one's appetites, and influencing one's friends to such effect, he maintains, are virtues (Book III:5, III:12, X:12).

5. Nordbakke and Schwanen (2013) identify ten combinations: (a) the utility approach, economics; (b) hedonic approach, psychology; (c) eudaimonic approach, psychology; (d) basic needs approach; (e) the resource approach (also known as the Scandinavian Level of Living approach); (f) the integral needs approach; (g) the capabilities approach; (h) the health-related quality of life (HRQoL) approach; (i) the lay views approach; and (j) the ecological perspective.

6. Is one long in the tooth—an expression literally referring to the recession of bone and gum from teeth and their increasingly elongated appearance? Or perhaps one is over the hill, past it, or a burden? Is one growing old gracefully or with some other mien? Either way, Phoenix and Grant (2009) suggest that the narratives and discursive practices of aging are imbued with power, and constitute and normalize what aging is, what appropriate physical activity is seen to be, what appropriate appearance and conduct are deemed.

7. This term refers to bodybuilders who do not take steroids.

8. In work funded by the Australian Research Council between 2008 and 2011, my colleagues and I found that rurality can be a confounding factor in terms of social disengagement among older adults, but can be ameliorated effectively by means of tailored community development activities and suitably sensitive provision of services (Walker et al. 2012, ARC Linkage Grant LP0882497).

9. Reinforcing observations that I made above about the constraints that different life contexts present, Umstattd and Hallam (2007) found that, for their survey population, regular exercise was associated with married white men of higher incomes and educational status, and for whom self-regulation, self-efficacy, and high outcome-expectancy values were present.

10. Cardiometabolic syndrome is a "constellation of maladaptive cardiovascular, renal, metabolic, prothrombotic, and inflammatory abnormalities . . . now recognized as a disease entity by the American Society of Endocrinology, National Cholesterol Education Program, and World Health Organization, among others" (Castro et al. 2003, 393).

7   The Undiscovered Country

In the British Museum is an installation entitled *Cradle to Grave* produced by 'Pharmacopoeia', a group comprising Susie Freeman, a textile artist; David Critchley, a video artist; and Dr. Liz Lee, a general practitioner. The work includes a fine, silver-grey net fabric, forty-two feet (thirteen meters) in length and one foot, eight inches (over half a meter) in width (Figure 7.1). Down each half-length are placed stories of men's and women's lives in captioned personal photographs, documents, and objects (Figure 7.2). Incorporated into the net are two lifetimes' supplies of over fourteen thousand prescribed pills, lozenges, tablets, and capsules—"the estimated average number prescribed to every person in Britain during their lifetime. This number does not include over-the-counter remedies, vitamins or other self-prescription pills" (The British Museum 2014, n.p.). In September 2011, as I drifted up and down the installation's length and mulled over the potential pages of this book, I knew *Cradle to Grave* could be an entry point to ask *how are we to understand the passage from liveliness to enfeeblement and dying into death, especially among the oldest-old? How are we to reflect on the geographies, mobilities, and rhythms of life at its close?* Here, then, I want to reflect on whether, how, and to what extent it is possible to flourish through this most challenging of prospects. Unexpectedly, my entry into this discussion has been via a route more circuitous than that characterizing other chapters; this is because, belatedly, I have been reminded that dying and death affect every stage of the life-course, and are the subjects of significant interdisciplinary scholarship (Davies and Park 2012), including among geographers. I have felt compelled to acknowledge all that in what follows.

"NO ONE GETS OUT OF HERE ALIVE"[1]

Avril Maddrell (2013, 503) reminds us that death "is an everyday reality" and that there is "considerable scope for cultural geographers to offer more insight to this universal and often life-shifting manifestation of absence". In chapter one, I referred to young Robin Walker from the *0 to 100 Project* and, accounting for his relative privilege, I suggested that if he should sit

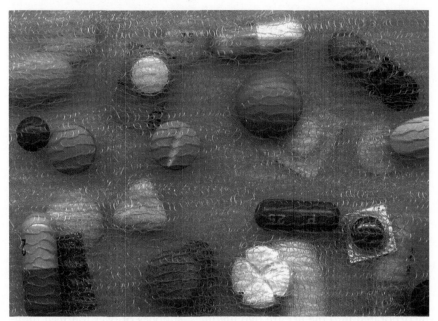

Figure 7.1   Knitted Pill Sampler for *Cradle to Grave* by Pharmacopoeia, 2003.
Photo: ©Pharmacopoeia. Courtesy of Pharmacopoeia (Susie Freeman, David Critchley and Dr. Liz Lee). Reproduced with permission. www.pharmacopoeia-art.net.

down to his hundredth birthday dinner in 2110, Robin's life experiences
might be composed of the most astonishing geographies, mobilities, and
rhythms. Globally, over the period since the 1950s, there has been a slow
increase in the likelihood that other babies, including those in marginalized 'elsewheres', will reach what has been called the fourth age—life as the
oldest-old. That likelihood starts in infancy: global and regional trend data
show universal declines in the infant mortality rate[2] from 1990 to 2012
(World Health Organization 2012). Rates nevertheless remain unacceptably
high: "The annual number of under-five deaths declined from 12.6 million
in 1990 to 6.6 million in 2012. Still, 18,000 children die every day . . .
Under-five mortality rates vary greatly by country: in Luxembourg, the
under-five mortality rate is just 2 per 1,000 live births; in Sierra Leone, it is
182 per 1,000" (UNICEF et al. 2004–2014, n.p.). More specifically still, in
Australia and Canada, one of which is the birthplace of the infant Robin
Walker, neonatal mortality rates were among the lowest in the world (UNI-
CEF et al. 2004–2014). Nevertheless—and this is my point—babies still die
in even the most privileged and advanced economies.[3,4]

In chapter two, I focused on the geographies, mobilities, and rhythms
of conception and gestation, purposefully ceasing that commentary at the
birth of a baby girl in Nilsson's (1982/1986) documentary, *The Miracle of*

Figure 7.2 Detail of *Cradle to Grave* art installation, 2003, by Pharmacopoeia: Early Years Woman's Narrative, 2003. Photo: ©Tom Lee. Courtesy of Pharmacopoeia (Susie Freeman, David Critchley and Dr. Liz Lee). Reproduced with permission. www.pharmacopoeia-art.net.

Life. Let me now return to the moment of birth by reference to a feature by Caroline Overington (2012) in the populist magazine *The Australian Women's Weekly*. The article is about a couple named Paul Murray and Siân Horstead and their infant Leo, who died thirty-four hours after his birth in Sydney on August 18 that year. In her account, Overington partly focuses on Deb De Wilde, the consultant obstetric social worker who attended the couple. In 1985, De Wilde and her own husband, Dr. Peter Barr—a physician specializing in the intensive care of neonates—made "an award-winning documentary, *Some Babies Die*, one of the first programs to gently lift the veil of mystery that surrounds the sudden loss of a child at birth or shortly afterwards" (Overington 2012, 64). In 1987, they then co-authored a book *Stillbirth and Newborn Death: Death and Life are the Same Mysteries*. For their works, both were awarded the Order of Australia.[5]

Overington documents De Wilde's description of her role as being to "encircle . . . parents, to keep them safe and try to slow things down for them, to allow them time to be with their baby, to begin creating memories . . . and to be guided in their actions not by fear, because we can be afraid of death, but by love" (63). She also refers to De Wilde's observation that over the length of her career, approaches of grief have changed from those emphasizing the need to forget or to embrace fatalism—epitomized by the phrase 'it was probably for the best'—to other methods privileging acts of remembering. For example, Overington describes how Paul and Siân spoke about having to rush to the hospital on August 18 because she was experiencing significant bleeding. Her obstetrician, Dr. Sean Seeho, then performed an emergency caesarean because he could find no trace of a fetal heartbeat after seventeen minutes. Then it was established that Leo had a vealmentous umbilical insertion in which the blood vessels in the cord ended six inches (fifteen centimeters) from the bulk of the placenta. A rare condition neither preventable nor detectable, in Leo's case it resulted in one of the vessels breaking, which meant that "Leo's lifeline basically bled out" (64), leaving him too compromised to survive long.

Overington describes how Paul and Siân recollected Leo's simple, wonderful, and utterly normal achievements during the hours in which he lived: eliminating, gesturing, gurgling, and watching. She and De Wilde also comment on the fact that, not so long ago, Leo may not have been held or named and, following his death, may have been rapidly removed from his mother and father. In this respect, research from the UK suggests that there remains much to learn about how to support those experiencing grief and loss from the death of a child, including insights about the importance of being able to choose the location where one's child is to face his or her final hours (Davies 2005). Overington also emphasizes the point that there is no term, in English at any rate, for "parents who have lost their baby" (63). Yet, because of the work done by people such as De Wilde, Paul's and Siân's loss was recognized; they were able to dress and hold Leo, "given time to tuck Leo into his casket", and enabled to create a "special service for him. Not a funeral, not a wake, but a celebration of the life of a child whose time on Earth was short but no less precious for that" (64). In this regard, the following observation on the importance of friendship to flourishing is noteworthy: "For the very presence of friends is pleasant both in good fortune and also in bad, since grief is lightened when friends sorrow with us. Hence one might ask whether they share as it were our burden, or—without that happening—their presence by its pleasantness, and the thought of their grieving with us, make our pain less" (Aristotle 350 BCE (b), Book IX:11).

The international comparative statistics on mortality noted above also make additional reference to mortality among young children. In this respect, in chapter three I pointed out that the members of this cohort are especially vulnerable to climate change. Jill Lawler et al. (2011, v) note the effects of climate change on "children's underdeveloped immune systems [which] put them at far greater risk of contracting ... diseases and succumbing to their complications". They point to the environment as a critically important influence on child health and morbidity, and underscore the relationship between child mortality and climate variability, especially in relation to the distribution of vector-borne diseases, extreme weather events, and heat stress. Malaria, for example, still accounts for over a million deaths per annum, mostly among children, and to it are attributed anemia, which affects growth and development; maternal morbidity and mortality; and low birth weights and neonatal deaths.

Perry Sheffield and Philip Landrigan (2011) emphasize the significance of these challenges in a review of primary studies, review articles, and organizational reports; for the year 2000 they attribute to climate change worldwide around one hundred and fifty thousand deaths, of which eighty-eight percent were among children. Noting the academic and scientific communities' focus on climate change and sea level rise, including amongst island states, they suggest that few studies focus on the health, impacts of such change on vulnerable communities, including children, and call for this gap to be redressed as a priority. Coming at such challenges from a different perspective, Sarah Strauss (2012, 372) notes that climate change is an "intensifier, which overlays but does not transcend the rest of the tests we face; it is therefore larger in scale and impact, perhaps, but not entirely separable from the many other environmental and cultural change problems" we already face. Certainly, many islanders already confront pressing problems related to public health, environmental pollution, land use, development pressures, and other challenges, and in some places, these are punctuated by changes in sea level (Figure 7.3). Strauss suggests that adaptation needs in one sphere are likely to map onto adaptation needs in other spheres, and she argues that it is in their intersections that attention should be focused. Given certain of her observations—that cultures always change, are the primary mechanisms for adaptation, are learned and shared, are integral to human *being*, and have their own logics and rules—it is not surprising that Strauss emphasizes the importance of storytelling to understand climate change vulnerability, adaptation, and resilience. Given that children's bodies also appear to be 'intensifiers', not least for risk, one wonders if significantly more attention directed to their care and flourishing might reap astonishing results.

It is salutary to consider the causes of death in the early decades of the life-course. Globally, the leading causes of death among children are pneumonia, pre-term birth complications, birth asphyxia, diarrhea, malaria, and malnutrition (World Health Organization 2014a); all of these point, too, to the extreme vulnerability of mothers. These trends are suggestive of the significant populations under twenty-five in developing economies. Among advanced economies, the story varies, especially if one acknowledges the existence of pronounced health inequalities affected by regional differentials in economic well-being. In the United States during the first year of life,

Figure 7.3   Over the 'borrow pits', Funafuti, Tuvalu. Source: Stratford, 2005.[6]

for example, mortality's chief causes are congenital, or implicate sudden infant death syndrome, low birth weight, and prematurity. From ages one to four, death is mainly caused by accidents, congenital disorders, or cancer. Between five and fourteen, accidents remain the chief cause of death, but cancer overtakes congenital disorders. From fifteen to twenty-four years of age, the three leading causes of death are accidents, homicide, and suicide (Medline Plus. US National Library of Medicine and the National Institutes of Health 2014). According to the UK Office for National Statistics (2006), in 2003, respiratory diseases, cancers, and injury, poisoning, or death from ingestion of noxious substances were the three chief categories of death among those aged from infancy to fourteen. The last of these was increasingly prevalent among those aged fifteen to twenty-nine. Analysis of several data sources from the Australian Bureau of Statistics (2007) establishes that the major causes of death among those aged to fourteen years were external causes, cancer, and neural diseases.

Of the thirty-six percent of all deaths attributable to external causes in Australia, fifteen percent were caused by traffic accidents; such accidents top the list in other advanced economies as well. Indeed, for Americans aged five to thirty-four, motor vehicle accidents are the leading cause of death, and account for the loss of eighteen thousand lives each year (United States Government. Department of Health and Human Services. Centers for Disease

Control and Prevention and National Center for Health Statistics 2013). Recall at this point that in chapter four I was concerned with ludic geographies, playfulness, and the question of how we might account for those who skate or trace. Recall, too, that the larger point was to contemplate the possibilities that generosity might bring to our capacity to flourish in settlements. Among the aforementioned thousands of road deaths in the United States, in 2011 an estimated forty-two skaters, and in 2012 an estimated twenty-eight skaters, were involved in fatal accidents, of which two injured in 2011 later died as a result of their injuries.[7] These deaths are reported in an Internet site, Skaters for Public Skateparks, based in Portland in Oregon, which serves as a national advocacy site for "safe skateboarding facilities around the globe" (Waters 2013, n.p.). In the 2012 report, reference is made to the kinds of analysis undertaken: of accidents involving long and short boards, locations of accidents, activities being undertaken, or of the use of protective gear by the skaters affected, among other variables. Although the report is limited to the US, additional reference is made to the incidence of deaths among skateboarders based in Australia, the UK, Canada, and South Africa. All of the reported deaths in the US occurred on roadways. One was a result from a collision with another skater, five involved falling on a roadway, three at crosswalks, three involved hit-and-run offences, eighteen were a result of being struck by a vehicle—a total of twenty-four implicated motor vehicles. Indeed, the "number one cause of death for skateboarders is bodily blunt force trauma caused by a collision with a vehicle, even one as small [sic] as a motorcycle" (Waters 2013, n.p.). All those who died were males ranging in age from eight to forty-two, the average age at death being twenty. Most deaths were in California, where street skating has largely been banned. It is noteworthy, then, that one of the final statements made at the bottom of the report on skating fatalities in 2012 is this: "For further information on Safe Skateboarding and Promoting Skateparks, visit our article 30 Reasons to Build a Skatepark in your Community!" (Waters 2013, n.p.).

Among the several ways in which these accounts may be read, it becomes apparent that advocacy does *not* include integrating skating as another form of commuting on roadways, and virtually nothing is written in either the scholarly or popular literature on such possibilities, although such matters do concern me in work forthcoming. Certainly, some of the skaters who lost their lives in 2011 and 2012 are likely to have been engaged in play for its own sake; the youngest, for example, was 'butt-boarding' down his driveway and simply and tragically went into oncoming traffic. In contrast, the eldest was skating to his car when he was struck. Described on the website as "a veteran skateboarder and long-time mentor, advocate and former shop owner in Chicago" (Waters 2013, n.p.), he was a mature adult and, using mixed modes of transport, was *commuting* at the time he was killed.

In chapter five, I considered the idea that commuting is a form of dwelling-in-motion susceptible to disruption, not in the least from acts of violence, and I will return to that point shortly. But movement *per se*, and commuting in particular, are vulnerable to the combined effects of speed and force, whether or not these are compounded by conditions either internal to commuters or external to them. Because it is not just rail passengers who dwell-in-motion in their commuting conduct (as was my concern in chapter five), here I comment briefly on those who drive or who are passengers in various kinds of motor vehicles. This consideration takes cognizance of observations such as that made by Mimi Sheller (2004, 221) that automobiles "are above all machines that move people . . . in many senses of the word"; as she points out, recent "approaches to the phenomenology of [automobility] have highlighted 'the driving body' as a set of social practices, embodied dispositions, and physical affordances". Thus, for Sheller, road trauma is a form of violent disruption to complex, emotional geographies of "automobility in which there are flows, circulations, distributions, intensifications and interferences of emotion between and among people, things and places" (223). Indeed, road trauma makes significant annual contributions to global and national statistics[8] measuring mortality, and it engenders a range of responses; not least is the transformation of the locations of fatal accidents to roadside shrines—places of commemoration in which the dead remain an absent-presence (Collins and Opie 2010; Meier, Frers, and Sigvardsdotter 2013).

It is noteworthy that, over the last several years, there has been a "considerable shift in the primary concern of US government agencies whose work it is to monitor, regulate, and govern transportation technologies. In short, rather than primarily focusing on avoiding accidents, the new mandate is to deter terrorists" (Packer 2008, 39). Also striking is that, in rail transport at any rate, suicide among adults is the main source of fatalities (UK Government. Department of Transport 2011). Thus, there is a painful irony in the attacks of 7/7, where three suicides on the London Underground deliberately resulted in the deaths of many others, not forgetting those on the double-decker bus that was the site of the fourth suicide.

In a paper on remembering that implicates those affected by these and other terror attacks, Matthew Allen and Stephen Brown (2011) examine death's absent-presence in the lives of survivors or families of the dead. They first describe a conversation with Susan Harrison, who had been a passenger on the Underground at the time one of the bombs was discharged between Kings Cross and Russell Square. Harrison lost her left leg in the explosion and suffered serious blast wounds. She has subsequently become an ambassador for The Healing Foundation UK (2004–2011, n.p.), which advances "the cause of people living with disfigurement and visible loss of function, by funding research into pioneering surgical and psychological healing techniques". With Harrison are sixteen others, two of whom were victims of terror attacks, one in 1999 and the other on 7/7. Allen and Brown then refer to two relatives of victims of terror attacks. One is Peter Fulham, the father of a young man, Matthew, who died in Egypt on July 23, 2005 when the resort town of Sharm-el-Sheikh in the Sinai was subjected to

Islamist attacks barely two weeks after the London bombings (British Broadcasting Corporation 2005b).[9] The other relative to whom Allen and Brown spoke is Esther Hyman, whose sister Miriam was killed in the last of the 7/7 attacks outside the British Medical Association headquarters.[10] According to Allen and Brown (2011, 312), Esther Hyman used the term *living memorial* to describe foundations that create "connections [with the dead] at the level of life rather than that of symbols". Examining this idea and its relationship to the notion of an affect economy, Allen and Brown (2011, 325) also demonstrate how acts of terror invoke various affective labors to celebrate life by means of the "conversion or 'salvation' of loss". In other words, they move us to seek different modes of conduct.

Finally, by way of moving to a consideration of the geographies, mobilities, and rhythms of dying and death among the oldest-old, let me return to the themes addressed in chapter six by reference to Stephen Katz (2000), who is among a number of scholars writing about the ways in which bodies become morally charged signifiers of the fruits of dressage, training, or conduct, and the desire to flourish. Among other things, his work points, first, to the rise in numbers of publications proposing that activity is an ethical idea(l) shaping gerontology and the ways in which we understand later life and, second, to renewed and recalibrated studies on disengagement "in connection with research on very old age or death and dying" (138). For me, Katz's work triggers varied musings about the etymological and substantive relationships between action and liveliness, and between liveliness and salvation (from which also derive the words salvage and salve; to save and to relieve). I make this association because, for a proportion of the middle-aged and elder populations, the cessation of movement is sudden. Such was the case with Jim Fixx, whose best-selling work, *The Complete Book of Running* (1977), I read in 1980, just four years before he died of a heart attack after his daily run on July 20, 1984, aged fifty-two. An obituary in the *New York Times* reported that Fixx's heart condition was hereditary—his own father succumbing to heart disease at forty-three. It emerges that by 1967, certain lifestyle choices had resulted in Fixx weighing two hundred and twenty pounds (ninety-nine kilograms) and smoking two packets of cigarettes a day. That year, he began running, developed a passion for it, lost sixty-one pounds (twenty-seven kilograms), and began to convert others to the benefits of running; this idea of conversion constituting a narrative of salvation that has interesting parallels with Ernestine Shepherd's story, outlined in chapter six. Despite there being no prior evidence of the severity of Fixx's condition, of the secret arrhythmias of his heart, one leading obituary noted that an autopsy revealed he "had died of a massive heart attack and that two of his coronary arteries were sufficiently blocked to warrant a bypass operation" (Gross 1984, n.p.). Indeed, the irony of a best-selling author of books on running dying of heart failure

> was not lost on Fred Lebow, the president of the New York Road Runners Club and the guiding light of the New York City Marathon. 'We

know running doesn't cause heart attacks and may, in fact, prolong life,' Mr Lebow said from Chicago, where he [was] attending a triathlon competition. 'What I'm concerned about now is all those people who talk about the danger of running. What does this prove to them? Sure, we have people dying in Central Park, one or two a year, while running. But I'm sure more people die on the golf course or watching the Yankees play baseball. Maybe if Jim Fixx didn't run, he'd have died five years ago.'

For others, the incapacitation is slower, sometimes insidious, and just as sad. In recent times it has been eloquently shared by those who experience such a demise, perhaps most prominently by the writer and commentator Christopher Hitchens (2012), who was diagnosed with cancer of the esophagus in 2010. He died in 2011 after documenting his battle with the disease and his understandings of life in such a light. On losing his capacity to speak in the manner in which he prided himself, and on which he had built an international reputation, Hitchens wrote the following, which is deeply reminiscent of Aristotle's evocations about a complete life being not simply about prosperity but connectedness:

> My chief consolation in this year of living dyingly has been the presence of friends. I can't eat or drink for pleasure any more so when they offer to come it's only for the blessed chance to talk. Some of these comrades can easily fill a hall with paying customers avid to hear them: They are talkers with whom it's a privilege just to keep up. Now at least I can do the listening for free . . . [while I] wait impatiently for a high dose of protons to be fired into my body at two-thirds the speed of light. What do I hope for? If not a cure, then remission. And what do I want back? In the most beautiful apposition of two of the simplest words in our language: the freedom of speech. (Hitchens 2012, 54–5)

This desire to document, to be heard, to be of account—and thus to leave a legacy—is also reflected in more populist tones in the case of Derek Miller (2011), a Canadian blogger, who was forty-one when he died from stage four metastatic colorectal cancer. In one blog from April 2, 2011, which Miller (2011, n.p.) entitled 'The time will come', he wrote an account of a shrinking world in which he could no longer walk the family dog, Lucy, and feared not being able to walk unassisted, or make food, or climb out of bed. I will not presume to speculate about what Miller found forbidding beyond that which is explicit in what he wrote: the loss of capacity to move is often attended by changed, diminished, or no ability to act for oneself. In a very real sense, what Miller described is a terminal point, a boundary, an end. Indeed, his final post, uploaded at his instruction on May 4 after his death, subsequently went viral. In it, Miller wrote this:

I haven't gone to a better place, or a worse one. I haven't gone any place, because Derek doesn't exist anymore. As soon as my body stopped functioning, and the neurons in my brain ceased firing, I made a remarkable transformation: from a living organism to a corpse, like a flower or a mouse that didn't make it through a particularly frosty night. The evidence is clear that once I died, it was over. (in *Daily Mail Reporter* 2011, n.p.)

"THE LIVING ARE GETTING RARER"

In the foregoing accounts, it is clear that across the life-course *being* involves senescence and the cessation of life. Present in such events are varied geographies, mobilities, and rhythms; different spatial and temporal scales; and diverse challenges to how we conduct ourselves in the face of great pain, grief, or loss—even where relief or peaceful resignation are intermixed with the experience. Organelles misfire, cells metastasize, flows are blocked, organs shut down, breath slows, blood stills, thoughts and actions dim and diminish, and then cease. Liveliness moves into apparent stillness, sometimes stealthily, or slowly, or with such rapidity we have no chance to adjust. Sometimes this transition is embedded in deeply unjust systems, processes, and structures, and seems alien—beyond reason, a heterotopia, or before someone's time—it is heterochronous. Sometimes, not least among the very old, this transition seems 'natural', which might also give it the appearance of somehow being 'just', although Gilleard and Higgs (2011) note that entry to the fourth age may involve deep abjection which, without the salve of intimate care from others, is anything but fair.

And once we are finally stilled, other spatial relations, other geographies, mobilities, and rhythms are invoked: in advanced economies, typically those might involve the movement of bodies from beds to mortuary gurneys, funeral homes, and on to sites of 'rest' such as cemeteries, which have their own particular geographies (Zelinsky 1994). Then, too, in the memories of those who remain, there is the instantiation of the dead's absent-presence and remobilization, including in commemoration rituals, memorialization, and ideas about the afterlife (Maddrell 2013). Assuredly, older participants in the *0 to 100 Project* understand these dynamics. Marlene Hasemer, at seventy-one, describes how "you look in your face, and you can't believe it's you. You sort of look and you say 'Oh, this is an old lady there'". John News takes a larger scan, remarking that he is "going to be spared global warming and runaway population growth and food shortages and all these terrible things that are coming on us, and I won't be suffering". Conversely, Aileene Taylor-Smith says of seventy-nine "It's a great age. I like it. I like it. I like it". Finally, Glyn Ellis observes of life at ninety-seven that the only thing he plans "to leave behind, I hope, is a good impression".

In this light it is interesting that, in *Rhinoceros*, playwright Eugene Ionesco (1960, 19) quips that there "are more dead people than living. And their numbers are increasing. The living are getting rarer". In short, even if one lives long enough to enter the ranks of the oldest-old, one will still succumb to something terminal, despite increasing life expectancy at birth (World Health Organization 2014b). Those predominant patterns themselves are instructive. The ten leading causes of death globally are, in order, ischemic heart disease, stroke, lower respiratory infections, chronic obstructive lung disease, diarrheal diseases, HIV/AIDS, trachea, bronchus, and lung cancer, diabetes mellitus, road trauma, and prematurity (World Health Organization 2013). Among US adults aged eighteen to sixty-five, causes of death in 2007 were malignant neoplasm, heart disease, unintentional injury (including road trauma), suicide, diabetes mellitus, cerebrovascular disease, chronic lower respiratory disease, liver disease, homicide, and HIV/AIDS (United States Government. Department of Health and Human Services. Centers for Disease Control and Prevention and National Center for Health Statistics. 2013). According to the Australian Bureau of Statistics (2013), intentional self-harm is the leading cause of death between fifteen and forty-four years of age in Australia. Thereafter, malignant neoplasms of the digestive organs and ischemic heart disease are the most prevalent causes of death between ages forty-five and eighty-four, the remaining eight of top ten causes include other forms of cancer, cerebrovascular disease, and diabetes mellitus. The situation changes somewhat among Australia's oldest-old aged eighty-five and over. Although ischemic heart and cerebrovascular diseases prevail, the third most common cause of death becomes organic (including symptomatic) mental disorders such as Alzheimer's disease. Influenza, pneumonia, renal failure, and other degenerative diseases of the nervous system also emerge as some of the top ten causes of death among the oldest-old.

Bluntly, if one lives long enough, one seems to wear out. How are we to understand such matters? Consider, as Kofman and Lebas (1996, 31) do, the ways in which Lefebvre thought of time as "projected onto space through measures", as then made uniform, and as emergent "in things and products . . . [that] give us a feeling of contentedness, constructing a rampart against the tragic and death". Indeed, they suggest that rhythmanalysis "is the means by which we understand the struggle against time within time itself . . . engraved [in] the passage from immaturity to maturity and the supreme difference, that of old age and death" (31). Important for any reflection on conduct and flourishing, they also point out the manner in which Lefebvre understood capitalism as "a system that is built upon contempt of the body and its life times" (31) that extends from the body of the person to cycles of nature and the cosmos (see also Lefebvre and Régulier 2004). In short, under certain forms of governing, and enmeshed in certain forms of conduct, no matter how long one lives there is a sense, according to Lefebvre, that people are constrained from developing in ways that might engender a sense of flourishing.

Also consider Sebastian Abrahamsson and Paul Simpson (2011, 331) and their examination of what constitutes the limits, boundaries, capacities, and thresholds of bodies, which, although limited in Euclidean space, also "extend beyond themselves . . . [and, among other things, can be] slowed down or held still". This extensible quality to embodiment is "relationally coupled with space and time" (332). Thus, the body is finite, deteriorates, and decomposes, but the end point is something "one can never experience . . . Rather, one is always exposed to it and does not cease coming up against it" (333). In other words, although we are assuredly among those who experience our dying (assuming we are cognizant), only others experience our deaths in ways that can be captured after we have departed. At the same time, one does not cease to confront different technologies and forms of conduct by which to challenge the body's finitude, among which are stem cell banking and bio-insurance, examples of a search to extend "life beyond its biological limits" (333). Nor does one fail to apprehend the quandary of when to foster life or let die—matters not just governmental but biopolitical, and thus deeply enmeshed in questions of conduct and the exercise of different forms of power (Dean 2002).

It is interesting, in this vein, that Abrahamsson and Simpson (2011) ask when are bodies considered properly whole, and what belongs to the body and to the world? So, for example, one might consider the biopolitical status of an elderly person with no living relatives who exists in a vegetative state in a nursing home bed and from whom might never come, for example, instructions that a properly whole patient could provide. Such persons are subject/subjugated to the biopower of medicine, through which the body is rendered legible (Foucault 1973). But where Foucault seems to have understood death as beyond medicine's reach, and by reference to over-comatose patients on life-support technologies,[11] Giorgio Agamben (1998, 104) argues that there exists something describable not as life but "death in motion".[12] Agamben also notes the existence of bodies that have become *faux vivant*—"kept alive by life-support systems" and that "would have the legal status of corpses but would maintain some of the characteristics of life for the sake of possible future transplants" (94). His larger point is that death is a political decision and, as such, it raises the specter of bare life. This is life exposed to death; not the qualified life characterized by a form of living particular to a group or individual, which Aristotle referred to as *bios* but life common to all living beings—*zoē*. We are become, Agamben (1998) suggests, *homo sacer*, those who, in Roman law, were outside (the protection of) the law and thus could be killed without being sacrificed (lost). This transformation raises a further danger: the question of what counts as life, as worth living, and what can be left to die? Along one trajectory, the resolution to that question teeters towards totalitarian ideas about those lives deemed unworthy of living. Along other pathways, it seems to me to be entangled in complex questions about salvation and sacrifice, and their relationships to ideas about a complete and flourishing life-course. Thus we return to the insoluble conundrum of the biopolitical status of an elderly orphan in a vegetative state in a nursing home bed.

These matters are part of a larger suite of questions that cut to the heart of how we conduct ourselves or empower others to do so on our behalf, either on individual terms or by means of informal and formal social contracts. Thus, in what follows, I want to consider the geographies, mobilities, and rhythms of dying as it has been experienced among the oldest-old in advanced economies, and do so by reference to home, aged care facilities, and high-needs and palliative care facilities. The (not inevitable, but not atypical) journey from one to the other does suggest two interrelated sliding scales—a diminution in geographical range of engagement and another diminution in agency (Martin and Vice 2012; Mitova 2012), and these are prevalent concerns in the literature. Certainly, the ways in which the 'shrinking' of life is experienced is also known to depend on other variables, such as social class, economic well-being, personal self-efficacy, location, and networks (Monroe, Oliviere, and Payne 2011).

For the very old, whether at home or in care, the composite process of living, becoming frail, and dying thus represents a significant choice and challenge—and I have more to add on the notion of frailty below. At the very least, this process implicates the configuration of space—its physical properties, its internal and external arrangements, and the ways in which they give effect to meaningful places. It also influences the geographical range of inhabitants, visitors, and caregivers; the reasons for and patterns of travel made by them; and their use or rejection of communication and virtual technologies to stretch engagement in the final stages of life—for example using telemedicine, accessing goods and services online, or staying in touch with distant family and friends. In the process, the motility and mobility of the dying and of those who intimately care for them may be profoundly affected. Without the elder's sanction, the body may transgress; without his or her *own* blessing or the blessings of *others*, the elder unwittingly may do the same by stepping over established boundaries or visibly shrinking within them. So, too, the lives of caregivers may be profoundly affected by the aging and dying in another; shifts in the rhythms of a life are inevitable as it winds down, and they ramify out and touch others.

Let me elaborate on these ideas starting 'at home'. There is widespread awareness that on psychological, emotional, and physical grounds the home is an ambivalent, uneasy, and sometimes dangerous space; nevertheless, the narrative constituting it as a haven remains potent. In part, this power exists because of the home's associations with "belonging, rootedness, memory and nostalgia" (Brickell 2012, 226). Evidence suggests that the elderly wish to remain living in their homes as long as possible, not least because that occupancy enables them to maintain existing networks and sometimes create new ones, for example by participating in events at seniors clubs (Aday, Kehoe, and Farney 2006; Walker et al. 2012).

Staying at home is both intrinsically and instrumentally important, and that is illustrated in a number of studies. For example, in one comparative study involving in-depth grounded research in Germany, Hungary, Latvia, Sweden, and the UK, Sixsmith et al. (2014) have established that among those very old people who are healthy, aging *in situ* is a negotiated process including a number of parameters. One of these is between the symbolic importance of maintaining homes to reflect and honor the worth of past lives and the practical importance of, for example, showing independence and capability by adapting internal fixtures and fittings to provide for appropriate living environs. In addition, Christine Milligan (2012) and Katherine Brickell (2012) point out that a widely-held aspiration to remain at home often leads to the boundaries between public and private spaces being blurred as elderly people make the transition to increasingly higher levels of care.

Sébastien Lord, Carol Després, and Thierry Ramadier (2011) describe this transition as one that need not reduce aging to the status of problem. They refer to a process, which in French is termed *déprise*, whereby daily life is reorganized to accommodate changed circumstances, without any insinuation of a 'downward spiral'. In my reading, this idea of *déprise* infers peaceable adaptation and quiescence (a calming quietude)—qualities that underpin a capacity to flourish. Those qualities are to be distinguished from acquiescence (a surrender close to capitulation), a tendency criticized by Lefebvre (1996) in his treatise on rights to the city and commentary on gratuitous gestures that are made to community participation in decisions about how to live. This idea of *déprise* stands in contrast to the evidence that people conflate the loss of productive roles, bodily decline, and confinement to the domestic sphere and consequently voice concern about the ways in which they come to be perceived as *frail* (see, for example, Mowl, Pain, and Talbot 2000).

Joanne Lynn and David Adamson (2003, 5) define frailty as "a fatal chronic condition in which all of the body's systems have little reserve and small upsets cause cascading health problems". Amanda Grenier's (2007) analysis of frailty is interesting in this light. She notes, for example, that the term's etymology "reveals an association with both physical and mental capacity (liable to break or be broken; subject to casualty; morally weak), and connects frailty with powerlessness and weakness (wanting in power), with impairment (subject to infirmities), and with an implication of blame (a fault arising from infirmity)" (430). Grenier also comments on the ways in which the assignment of frailty to the elderly influences assessments made by home-care professionals, and is used to determine access to a range of services, and suggests, after Foucault (1972), that frailty becomes a dividing and subjugating practice.

Less apocalyptic than Lynn and Adamson's (2003) definition of frailty is another provided by Nicholson and Hockley (2011): a state of in-between living and dying. Like Lord et al. (2011), they understand becoming frail as a *necessary* process if one is to mourn the loss of all things, including one-self, and then be able to "invest . . . in people and things that will outlive you" (Nicholson and Hockley 2011, 103). Yet again, there are strains of the dance between sacrifice and salvation here. In Nicholson and Hockley's assessment, this becoming-frail is best aided by "the presence of people to engage with stories, recognize and value daily rituals that anchor experience and facilitate creative connections [which] is vital to retain capacity, quality of life and the natural development into dying" (103–4). In my reading, such observations imply clear connections between Aristotle's idea of a virtuous and complete life and modern commentators' important and well-placed concerns about how to provide space for the dying to let go, and thus to flourish in their final days:

> For there is required, as we said, not only complete virtue but also a complete life, since many changes occur in life, and all manner of chances . . . Must no one at all, then, be called happy while he lives; must we, as Solon says, see the end? Even if we are to lay down this doctrine, is it also the case that a man is happy when he is dead? . . . that one can then safely call a man blessed as being at last beyond evils and misfortunes . . . for both evil and good are thought to exist for a dead man, as much as for one who is alive but not aware of them; e.g. honours and dishonours and the good or bad fortunes of children and in general of descendants. And this also presents a problem; for though a man has lived happily up to old age and has had a death worthy of his life, many reverses may befall his descendants—some of them may be good and attain the life they deserve, while with others the opposite may be the case; and clearly too the degrees of relationship between them and their ancestors may vary indefinitely. It would be odd, then, if the dead man were to share in these changes and become at one time happy, at another wretched; while it would also be odd if the fortunes of the descendants did not for some time have some effect on the happiness of their ancestors. (Aristotle 305 BCE(b), Book I:9–10)

Thus it is important to account for the ambivalence and uncertainty that attach to becoming old, or indeed very old, at home, because they do not dull a widespread desire to live *and die* where one dwells. This desire is tempered by instances, among the homebound elderly, of uncertainty spilling over to fear—and they are not alone in this. According to P.S. Fry (1990), a factor analysis of one hundred and seventy-eight very elderly participants based in Canada identifies the existence of three categories of concern: fear of pain, suffering, and sensory loss; fear of risks to personal safety—especially around the time of death; and threats to self-esteem and dignity during the process of dying itself. Entangled in these broad categories of fear are three specific and pronounced concerns—dying alone, being forgotten, and fear of what lies beyond death. Fry's study is, however, without *nuanced* analysis of what boundedness at home means; the term is used merely as *mise en scène*. Nearly twenty years later, Alison Kenner (2008) provides useful analysis showing how the meaning of boundedness has changed as the home has become, in this instance, a space of surveillance of elderly people with dementia who have a propensity for wandering. Sensors placed on the body and, or around, the home, are intended to track ambulation and activities of those being monitored, and are used both in homes and special accommodation facilities that serve as 'home'. Such technologies augment the panoply of other tools by which the life-li-ness or demise of individuals is registered remotely and, according to Kenner, they bring into sharp focus the political economy of aging, and the deontic challenges faced by caregivers when asking how best to keep the frail or the dying safe and at ease without robbing them of their dignity.

Clearly, becoming frail and dying affect those who die and those who care for them (Andersson et al. 2010; Froggatt, Hockley, and Parker 2010; Reich, Signorell, and Busato 2013; Renz et al. 2013), and dying at home makes particular demands on caregivers. In one study of ten caregivers interviewed six to twelve months after their family member had died, Ida Carlander et al. (2011, 1097) established the development of what they have termed a *modified self* that arises from the experience of care until death and centered on "challenged ideals, stretched limits and interdependency". Caregivers may experience a range of very stressful physical, psychological, and emotional effects, although the overwhelming evidence is that they feel "a great overall satisfaction with the caring experience" (1098). Indeed, among caregivers in Sweden, at least being "*next of kin to an old person in the last phase of life meant being a devoted companion during the transition to the inevitable end*" (Andersson et al. 2010, 20; original emphasis). Nevertheless, their ideals about how to conduct themselves are often tested, and they report that sometimes they reacted in ways they did not recognize in themselves. Their capacities are often stretched to, and beyond, limits they feel they can cope with, especially when care requires intimate interactions with bodily functions. They also describe the ways in which the needs of the dying take priority over the needs of others, sometimes for extended periods, with the effect that lives become deeply intermingled and interdependent, in ways Carlander and her colleagues describe as affecting the very rhythms of daily life.

At the same time, Mary O'Brien and Barbara Jack (2010) note that dying at home occurs among only approximately twenty percent of people of all ages in places such as Australia, New Zealand, the US, and the UK. This trend differs from a century ago, and was precipitated by widespread access to healthcare, and what O'Brien and Jack term the medicalization and institutionalization of death. Indeed, in a scathing work on the expropriation of health, Ivan Illich (1976) has suggested that modern medicine is in epidemic proportions, that life has been thoroughly medicalized, that dying is the ultimate form of consumer resistance, and that technical death has triumphed over natural dying. Thus, there are high levels of concern about dying in institutional care under deeply interventionist conditions. Nevertheless, for varied reasons, many people who are dying do not feel in a position to stay at home, and many caregivers, if there are any, are not in a position to assist them to do so. Simultaneously, health organizations and governments find it difficult to provide sufficient palliative and after-hours care, adequate funds, and enough training of professionals to better aid those wishing to die at home. Where such services are possible, John Rosenberg (2011, 23) argues from a basis of experience as a community palliative care clinician that "an acknowledgement of the home as a place of the person's dying was essential in establishing a relationship that was both therapeutic and interpersonal . . . in this place, and indeed, elsewhere in settings of care". And so it is that still-significant numbers of the oldest-old die in such 'elsewheres'.

What of the geographies, mobilities, and rhythms that attend institutionalized care? It appears that many people confronted with mortal illnesses or conditions express a strong desire to die at home. Nevertheless, in the United States, for example, a preponderance of the elderly still die in institutionalized care settings—residential aged care facilities or nursing homes, hospitals, and hospices (United States Government. Department of Health and Human Services. Centers for Disease Control and Prevention and National Center for Health Statistics 2010). Analysis of longitudinal data suggests that more people are dying at home, something near twenty-five percent in 2007 as distinct from the fifteen percent in 1989, and many of them are likely to be under sixty-five years of age. Even so, just over a third of people die in hospitals as in-patients, and nearly thirty percent die in nursing homes, the vast majority of them over sixty-five. The US Department of Health and Human Services staff acknowledges that many factors influence the place of death: cultural mores, social contexts, economic circumstances, and government policies chief among them.

Studies in other settings reach similar conclusions. Reporting on the care of very elderly and dying patients in a southern municipality of Sweden, Magdalena Andersson, Ingalill Hallberg, and Anna-Karin Edberg (2008) focus on the qualitative experiences of patients aged seventy-eight to one hundred years who were close to death at the time of the interview. Shared narratives among seven men (three of them in care) and ten women (five in care), elicited a number of themes in both home and care settings: retaining dignity; enjoying little things; feeling 'at home' (in bed, with routines, in rooms); being safe in the hands of others—especially caregivers in institutional settings; trying to adjust (to past, present, and imminent) loss; feeling important to other people; and remaining alive while facing death. At the same time, in both settings it was possible to feel great loneliness, and people in care reported sometimes not feeling at home because of their perceived approach to them of medical staff or because of a sense of having to 'fit in' and 'behave' in particular ways. This notion of 'fitting in' has, as a corollary, pronounced risk or fear of being infantilized, a matter that Audrey Anton (2012) argues is morally wrong on the basis of deontic intuitions about respecting our elders.

Other work by Katherine Froggatt and colleagues (2010) summarizing a number of studies in the UK, Australia, and Canada suggests that care homes are contested places in which the life-worlds of residents/patients, families, and caregivers are experienced as contested states. They describe the care home setting as residential, collective, geared to provide for chronic (independent), acute (dependent), and palliative (comforting end-of-life) care, and involving family, other trusted parties who are not professionals, and professionals. They also note that in the context of acute needs, the care home becomes the substitute for a hospital; in palliative care it is a hospice. Through such settings, people circulate in challenging geographies that involve complex patterns of mobility, and that must account for diverse embodied, diurnal, or governance rhythms such as might exemplify *waiting* lists. As Froggatt and her colleagues (2010, 264) suggest, the "context is a complex one crossing many boundaries". Moreover, as they emphasize, if or when residence is taken up in a care home, an elder's frailty has become established and is often made more complex by comorbidities such as advanced dementia. These conditions make for additional challenges to families and staff as well. First, no one involved in the care of the very old is uniform or homogeneous in need, approach, or context, and yet resource constraints often mean that residents and patients live in "difficult circumstances and [experience] poor care . . . [and] variable access to appropriate services" (265). At the same time, families may feel deeply confronted by conflicting expectations—the elder's, their own, and those of staff.

The last of these—staff who are often middle-aged women and increasingly from migrant backgrounds—also experience poor and physically testing working conditions, report feeling undervalued, receive few rewards and little recognition, and function with limited resources and minimal training. Compounding these conditions, Froggatt et al. (2010, 266) report that those involved in supported care work in a context of profound risk aversion and economic rationalism characterized by significant funding differentials and high levels of regulation focused on meeting "legislative standards and [securing] ongoing financial reimbursement". The conclusion reached about these varied forms of contestation is noteworthy, the authors suggesting that it reflects wider dynamics between life-worlds and systems that reflect a culture intent on avoiding death. Indeed, in what they term a *living-dying interval*, the authors discern a lack of openness around dying and death, the net effects of which are that "residents are not consulted and decisions are made about their care by health professionals that lead to experiences of care, for example in hospitals, that do not reflect resident's preferences" (267). Certainly, the efficacy of advance care directives[13] has been brought under question as a result of these trends. Ultimately, Froggatt and her colleagues intimate that there is real need for care based on relationships rather than break-even policies or profit motives. This recommendation accords with Hollie Mann's (2012) thoughtful paper on ancient virtues and contemporary practices in relation to embodied care. There, she critically rereads Aristotle, interpretations of his work, and the work of feminist care ethicists such as Joan Tronto (1987, 1993) and Virginia Held (1989), following whom she argues not only "then, does care not stand opposed to just political life. A just political life might not be possible without an ethic of care" (Mann 2012, 195). Drawing on Aristotle's use of *philia*, which Mann (2012, 198) describes as "the mutually acknowledged and reciprocal exchange of goodwill and affection that exists between people who share an interest in each other on the basis of pleasure, usefulness, or virtue" she also notes that Aristotle included in such understandings of love both bodily care and the *explicit* care of the elderly. New readings of Aristotle[14] allow several ideas to cohere around the functions of *oikia* (family), *techne*, and *praxis*, and around "the openings he creates for thinking about the centrality of giving care (not merely receiving it) to human flourishing" (198). Indeed, one of Aristotle's theses was that we are "more complete when we actively do well by others with whom we share a moral life and political constitution" (204). Acknowledging that Aristotle gave little credence or value to manual labor as virtuous, her point is that we should be asking "whether or not Aristotle *should* defend embodied care . . . especially in light of his views on human flourishing, activity, and doing well by others" (207; emphasis added). Given the profound vulnerability of the dying body, and the exposure to life's contingency of those who experience dying and those who care for them, Mann stresses the importance of realizing *simultaneously* that we are an animal and that, for us, caregiving enables a "becoming human". Mann's (2012, 214) conclusion is also testing for what it asks of us, whether one interprets her reference 'to live with' as literal or not:

> Participating in communities of care is both a way of living excellently and an important method for determining how to best achieve the excellence that is caregiving . . . But if care truly is a virtue, then we need to better understand why we ought to live with [the dying] rather than merely secure the material conditions that make their care (by someone else) possible. We need to come to see how living with those who need our care in order to live well can enrich our own lives.

THE LAST CHAPTER

Whether one dies at home or in institutionalized care, is conscious and alert and present in one's body to the end, or apparently is 'missing in action'— indeed no matter the context of a death—Christopher Cowley (2013, 3) argues that it is critically important to the dying to know that the 'last chapter of the story' is written as well as possible; the alternative is "the source of the peculiar vulnerability to humiliation suffered by the elderly". Cowley's focus is on those who have reached a point of understanding that it is time to prepare for death, and who seek to flourish in so doing by reference to the whole of a life that has been lived. In this sense, *eudaimonia* is retrospective; yet its fulfillment is anticipatory, deliberate—a form of conduct, and intended to be a legacy. Real lives so understood are much larger and more complex, Cowley argues, than those which typify explanations of the life-course based on thin liberalism and physiological deficit. Of note, Cowley's account renders the life-course a geography characterized by moving experiences—encounters that both arouse affect and whose details shift in their telling and retelling. He writes that the "precise significance of an event . . . will depend on my precise location in my life's journey, and on the context of that location, since I will be looking at the event from that location [which] . . . has a broad scope to include all my projects, relationships, concerns of that moment, as well as the route by which I got to this location . . . there will remain wide room for a shifting significance of the same event . . . [it] is arbitrary" (5–6). In my reading, such arbitrary qualities render no event less meaningful to the individual, but they do infer that the scripting of the last chapter constitutes an *intervallum*—a space in which to write of a life well lived, a death well prepared for, a legacy (Goggins 2012); and it also establishes a *diastema*—a necessary and inevitable distancing in order to depart.

The sort of reflection to which Cowley refers is, in fact, exemplified in the writing of Stanley Leavy. A widower, psychoanalyst, and retired clinical professor of psychiatry at Yale University, Leavy (2011, 701) elaborates on the conceptualization and experience of death through the life-course, and then focuses on two questions: "what about my own death? How can I face it?". Defining "manifest death" as in correspondence "with the end of all subjective experience" (703), Leavy is matter-of-fact in arguing that the disintegration of body and brain means "the end of experience for the dead. The body does not disappear [immediately], but the true person, the self, does . . . We think too little about what dying is . . . All that I can be aware of, in writing at this moment of conscious experience, is the statistical certainty of my biological death" (703–4). He later muses that in losing his life, life also loses him: we become the "screen on which the coming event" of dying is projected, and for the elderly the "inevitability and proximity of death" is particularly poignant in this regard (705). And here, Leavy's analysis turns to the idea of flourishing, to seizing the day for acts of greater good, and to prudential reasoning of a kind that echoes back to the aforementioned insights provided by Dean (2002), Rose (2007), or O'Malley (2000) (and see also Scruton (2012) on timely death and the virtues it entails). Of critical importance, however, Leavy notes that such reasoning leaves little room for the possibility that "old age might have something good to offer in itself" (706), that it might be "unique, new, fresh, and that its benefits exist not in spite of physical and mental limitations, but joined with them" (708). This work is revelatory, insisting that we rethink the meaning of 'old', and reinscribe in the present absented understandings of it as experienced, nourishing, and strengthened (OED). For Leavy, this state of being must also account for the interconnectedness of a life with others, and he reveals both the deep sense of loss he feels for his wife, and a function of loving to protect us "from the fear of death because it relegates to surviving others the continuity that we are forced to give up by dying" (713).

What, then, might the dying want and require? Assuredly, they seek time and space in which to come to peace about the prospect of their demise; situations in which their wishes are respected; their dignity, identity, cultural practices, and humanity observed; trust in caregivers well-placed and honored; suffering minimized and treatment agreed; finances in order; families cared for; partings and goodbyes made; conflicts resolved; humor retained; companionship assured; and funerals planned (Aziz, Miller, and Curtis 2012; Steinhauser et al. 2000). In short, as John Cottingham (2012) suggests in his analysis of the question of aging and Aristotelian views therein, we seek affirmation of *entelechy*—the fulfillment of our potential, which is secured by growing old and perhaps goes some way to explaining why death before great age might somehow seem unjust.

NOTES

1. This lyric is from a song by The Doors, *Five to One*, by Jon Densmore, Jim Morrison, Robbie Kreiger, and Ray Manzarek.
2. According to internationally agreed definitions, infant mortality rate (IMR) is a measure of the probability of dying between birth and one year of age, and is expressed per thousand live births. Neonatal mortality rate (NMR) refers to the probability of dying between birth and twenty-eight days per thousand live births.
3. Perinatal cause of death from 2001 to 2010 is enumerated, for example, by the Australian Bureau of Statistics and includes analysis of conditions originating in the perinatal period that affect fetuses or newborns (Australian Bureau of Statistics 2012, n.p.).
4. One of the many paradoxes of our ways of living and dying is illustrated not by the death of a child, but the death of a partner prior to the conception of a child. The most recent instantiation that I can find was in Adelaide, South Australia in 2013. There, a woman whose husband died in a car accident in 2011 sought and was granted permission by the Supreme Court to access sperm from him that had been in storage for IVF treatment, but was required to undertake the treatment in the Australian Capital Territory because of South Australian laws, and was also required to consult with the parents of her deceased partner (Founten 2013).
5. "The Order of Australia, instituted by Her Majesty The Queen on 14 February 1975, was established as 'an Australian society of honour for the purpose of according recognition to Australian citizens and other persons for achievement or for meritorious service'" (The Order of Australia Association 2014, n.p.).
6. The borrow pits are large excavation sites dating back to World War II (Chambers and Chambers 2007), at which time allied forces used coral sands to construct an airstrip capable of landing and deploying B52 bombers. Reconstruction works were never undertaken, and the pits regularly flood or are subject to seepage from seawater, fill, or are filled with solid waste, including hazardous materials, and are a serious composite problem.

7. There are few statistics on what one might contingently term parkour fatalities, because it is unlikely that they would be enumerated as such. One Internet site has released the results of a lay survey done in 2012 among *traceurs* in the US and, not surprisingly, the most prevalent injuries among 239 respondents were to elbows, wrists, hands, knees, and ankles; involved sprains and contusions. These resulted from falls during precision vaults, from miscalculating heights, or from freerunning maneuvers (Parkour Conditioning 2014). Google searches do produce links to reports of *traceur* deaths, mostly involving falls from heights, but statistical analysis is largely absent.

8. In the United States, 44,074 people died in road accidents in 1999, of which nearly half were passenger car occupants and over ten percent were pedestrians struck by motor vehicles. Most, expectedly, were adults. By 2011, the total had declined to 34,360, of which just over a third were passenger car occupants. Almost thirteen percent were pedestrians struck by motor vehicles (United States Department of Transportation 2013). In the United Kingdom, in 2012 some 195,723 casualties were reported to police, and of those 1,754 people died; this number of fatalities included a decrease across categories except for pedal cyclists, among whom fatalities increased from one hundred and seven in 2011 to one hundred and eighteen in 2012 (Lloyd et al. 2013). Scrutiny of various statistics reveals similar stories in other advanced economies such as Australia (Australian Government. Department of Infrastructure and Regional Development. Bureau of Infrastructure 2013), Canada (Government of Canada. Transport Canada 2010), and New Zealand (New Zealand Government. Ministry of Transport 2014).

9. After the event, the family established a foundation in Matthew's name that, between 2008 and 2012 provided bursaries for young musicians and funds for an annual 'proms in the park'. After 2012, funds were insufficient to maintain the bursaries and the local council took on the annual event, the fundraising page noting the following: "We are no longer fund raising for this cause. As a charity we have found it more and more difficult to raise funds, not because people are not willing to support us, but because we feel that there are other more important charities that need supporting like Cancer and other health related charities" (The Matthew Fulham Foundation 2012, n.p.).

10. The Miriam Hyman Memorial Trust was founded in 2005 and its focus has been on providing technical and expert support for eye clinics such as the L.V. Prasad Eye Institute in Bhubaneswar in India, although new forms of fundraising are listed on the news and events page for January 2014 (Miriam Hyman Memorial Trust 2014).

11. Specifically, Agamben (1998) refers to twenty-one year old Karen Ann Quinlan, who was in a comatose state for a lengthy period after accidentally combining a number of substances at a party. When the decision to take her off life-support was made, she continued breathing and died of pneumonia four years after that time.

12. This phrase is reminiscent of the idea that our circuits of vitality have become "stabilized, frozen, banked, stored, accumulated, exchanged, traded across time, across space, across organs and species, across diverse contexts and enterprises, in the service of bioeconomic objectives" (Rose 2007, 38).

13. Advance care directives, sometimes known as living wills, are written statements that enable individuals to document their "wishes for care and treatment should the time come" when they are unable directly to communicate those wishes. They often rely on persons entrusted with powers of enduring guardianship and/or power of attorney (The Advance Care Directive Association Inc. 2014). Directives are considered legally binding if signed when the individual is deemed competent, are supported by significant numbers

of medical professionals, and yet, "where the patient was likely to survive if given a particular treatment, most participants [in an UK study of geriatricians] expressed discomfort in following a written request to withhold this" (Bond and Lowton 2011, 452). Emergencies may also mean that the existence of directives is not known until after treatments are provided.

14. Such readings also include those by Cheshire Calhoun (2004) and Marcia Homiak (2004) to whom I turn in more detail in chapter eight.

8 Space to Flourish

I began this work with the unremarkable observation that our lives are increasingly mobile, and noted that this trend is, nevertheless, intrinsically interesting and of instrumental worth to people working in diverse fields concerned with what it means to be in the world. In passing, I observed that mobile lives influence and, in turn, are influenced by a range of technologies, among which are iPads, apps, and the Internet. By means of such technologies, Nicholson (2011) was able to create and deliver the *0 to 100 Project*, an arts initiative that sensitively interprets people's perceptions of their place in the life-course, their varied experiences, and diversity. I acknowledged, too, that where one is located, literally and metaphorically, confounds our experiences of living, giving effect to such things as life expectancy. The *0 to 100 Project* was deployed through the pages of this book as a heuristic device that has enabled me to think about our journeys over the life-course and the adventures that we have in its intervals. I used this idea of the interval (and thought of it in terms of heterotopia and heterochrony) to signify the existence of gaps, distances, and pauses—spaces in which alternatives might be thinkable, in which potential might be explored, and in which moments of radical uncertainty might not only test, but delight or dismay.

I introduced two key threads that were then woven through the work. The first concerned the conduct of conduct—ways of governing ourselves, governing others, and being governed in spatial and place-based relations; in our movements and capacity to be motile; and in the rhythms of moments, days, weeks, or decades. The second concerned the idea of flourishing as both a means and an end. Linking the examination of conduct and flourishing allowed me to ask *as we move through our lives how do we conduct or govern ourselves and each other, and in doing so how are we to constitute the spaces and flows for a life that is flourishing?* I arrived at these questions by responding to four invitations in the literature: Cresswell's (2010b) suggestion that we take up a more finely developed politics of mobility; Anderson and Smith's (2001) proposition that we account more often and more appropriately for emotional geographies; Ellis, Adams, and Bochner's

(2011) proposal that we better value and use auto-ethnography in our work; and Edensor's (2014) call for more in terms of studies of rhythms.

I sought to respond to this fourfold agenda by applying a number of overlapping theoretical frameworks to a series of case studies I am drawn to. By means of an extensive review of the literature, I stressed the importance of thinking about the geographies, mobilities, and rhythms that constitute the life-course and, in particular, I emphasized the importance to my thinking of multiple works by Lefebvre, Foucault, Aristotle, and others who have worked with, built on, and critiqued their ideas. As part of the initial mapping of terrain on which the book was founded, I made specific reference to ideas about the ethics of mobility promulgated by Bergmann and Sager's (2008), to Wright's (2010, 2013) ideas about the possibility of real utopias, Cloke's (2002) call for geographically sensitive ethics and an ethically sensitive geography, and Tuan's (1999) questions about how we constitute what crushes us and might sustain us. These, then, were the weft threads that I threaded through five cases that move, interest, and concern me.

In chapter two, I began with Nash's (2005) ideas about places of origin and ended with Northcott's (2006) concerns that we have created new geographies of exile by displacing the embryo from the uterus. My intention was to work out from Nash's ideas about the geographies of relatedness and ask how one might think through the geographies, mobilities, and rhythms involved in bringing to life or life-li-ness in ways that may or may not lead to birth. This question prompted considered engagement with a number of narratives based in other disciplines that nevertheless were shown to be deeply geographical, to mobilize all kinds of events, and to be based on all sorts of rhythms. Among such works were those by Streeter, England (1983), and other embryologists, medical and other scientists, ethicists, feminist theorists, and others such as Rose (2007) and Northcott (2006) who, in markedly different ways, respond to questions about bringing to life. I attempted to account for these origin stories, providing a genealogy of embryology and attending to critiques of it, of stem cell research, and of related science, and pondering—without resolution, of course—not only what is life, but how, where, and when is life? Then moving to consider the labors performed by photographers and filmmakers such as Nilsson, I noted the power of their choices in picturing life, before moving to consider the importance of work by feminist scholars and ethicists. Such work revealed, as Haraway (1997) and Franklin (2006, 168) noted, an "embryo-strewn world", diverse visual and other economies of the unborn, the use of abortuses as metonymy for life-li-ness removed from the body, and the development of technologies that take cellular and molecular matter and transform it for diverse ends. These revelations pointed to the production and consumption of heterotopias and heterochronies that are constitutive of what Rose (2007) refers to as the politics of life itself. Demonstrably, I came to understand that there are powerfully influential geographies, mobilities, and rhythms at work here. Arguably, we need to continue to account for them as profoundly testing debates continue to arise about places of origins and possible exile, about how we are to map our conduct in the new terrain of emergent biotechnologies, and about the ways in which our decisions—or indeed our acquiescence (as Lefebvre and Régulier [2004] would have it)—will give effect to our collective capacity to flourish.

Chapter three provided opportunities to consider another kind of fluid terrain that might involve the loss of habitable lands—especially islands—if predictions about anthropogenic climate change, sea level rise, and inundation hold true. What then for island children, among others? How do we account for questions about displacement, evacuation, relocation, or migration on one hand, and about citizenship, sovereignty, nationhood, and resilience, on the other? I started to examine that conundrum by reference to McTaggart's *The Storm*, and its powerful and varied geographies, mobilities, and rhythms of coastal, peninsular, or island life. I noted how the painting is emblematic of changes we are likely to confront, depicting as it does people in distress in a boat in a storm, and people looking on from the shore—I assume with a view to help. Some are children, who bear witness to the unfolding events before them. My own bearing witness to the presence in the painting of the children prompted consideration of Skelton's (2010) observation that young people are not political subjects in waiting, which requires that we revise Aristotle's understandings of their place in society. I also referred to Lee and Motzkau's (2011) work, which positions children not as human becomings, but rather as full citizens, and which invites us to think about the impact of our conduct on children's capacities—minimally to function and optimally to flourish. These ideas, it will be recalled, also inform the rise and development of children's geographies, and its proponents' shared assertion of the intrinsic worth of engaging with young people. Drawing on works by Bell (2008) and Thomson (2007), I emphasized the merits of rights-based approaches and the need to create more space for thinking *with* children, and not just for them. In this respect, and given the complex geographies, mobilities, and rhythms that are likely to attend changing conceptualizations of citizenship under conditions of climate change, it was useful to be able to work out from Golombek's (2006) commentary on *jus solis, jus sanguinis*, naturalization, and majority, and emphasize the importance of recognizing children as actively engaging in life and aspiring to be treated in ways that enable them to flourish and be heard. These reflections were enriched by Gardiner's (2010b) understanding of climate change as a perfect moral storm that immobilizes, an intergenerational storm characterized by backloading, and a theoretical storm that requires new ontologies, epistemologies, axiology, and methods. Indeed, Gardiner's concerns about how we understand our legacies and simultaneously deal with pressing problems in the present underscore what it might mean to construct a complete life. Those considerations led me to ask what the future holds for islands and island peoples, and—as others have—to caution that we must take care to neither underestimate their vulnerabilities nor disable them in the process of advancing governmental and legal solutions to the possibility of displacement. Among such solutions are truly useful ideas such as that proposed by Burkett (2011) for the creation of Nations *Ex-Situ*. Finally, I brought all these considerations to a focal point by reference to *A Map of a Dream of the Future*, a project which invited children to think about their political views using story, artwork, and performance, and then mapped those views using Political Compass and a multimedia arts installation. In that project, we came to understand that the children with whom we engaged had clear and thoughtful understandings of climate change and migration; diverse views on the roles of governments, individuals, and communities; and the capacity to distinguish between displacement, evacuation, relocation, and migration. Of real merit, we were also able to discern that the children sought novel solutions to the various scenarios we put before them, and wished to constitute real and figurative spaces of hope for the future.

The distance or interval between childhood and adolescence seems, in retrospect, to be but a moment—but of course its living may seem to take so long and, in the course of that passage of time, young people are often marginalized and sometimes vilified, not in the least in our cities and settlements. In chapter four, then, I made early reference to work by Child (2007, 2009, 2010), who has invited us to reengage with play and learn from the joy that young people may experience by simple acts of engagement with the geographies, mobilities, and rhythms of urban life. I focused then on skating and parkour as two key practices of young people, and juxtaposed these practices to walking, and to the scholarship on walking. I have been interested in how skaters and *traceurs* use space, contour space, and are themselves contoured in the process, and my reflections have led me to consider what it might mean to see them differently, and look to ludic geographies to enrich our lives. Those considerations prompted other reflections that informed the intention of the chapter—not least the idea that there are rights to the city that should be claimed in order to flourish (and that there is need to account, too, for our responsibilities; in the terms provided by parkour, to leave no trace of negativity). Preceding those thoughts were deliberations about how teenagers and young adults are defined in law and policy; about the ways in which they are, and create, a constellation of temporary coherence, in Massey's (1998) terms; and about the volume of literature that seeks to account for their geographies, mobilities, and rhythms. In this respect, I made reference to a number of works, among which was featured Jeffrey's reports (2010, 2012, 2013) on the range of structural issues affecting young people, the normative and teleological assumptions involved in thinking about life stages and how what it means to be of a given age varies over time and space (is heterotopian and heterochronous), and the ways in which young people enact multiple forms of agency in spite of stereotypical constructions of them as, for example, heroes or zeroes. Indeed, as Jeffrey noted, young people, too, navigate plural intersections of power and of life from the micro-scale to the global scale. This scaffolding in place, I then moved to consider in more detail the question of playfulness and the ways in which we have developed geographical and geopolitical understandings of playfulness. I gave particular emphasis to Woodyer's (2012) summation of debates on play (see also Aitken's [2001] commentary on thick play and the ways in which it gets squeezed to the point that imagination and potential are constrained). I also underscored the utility of Woodyer's reflections on ludic geographies, to which I will return shortly. Reference was made to de Certeau (1984) and the politics of walking the city as a distinctive and meaningful act simultaneously rhythmic, mobile, geographical, and to Debord (1956, 1967) and the Situationists, and their ideas about *dérive* and *détournement*. Consideration of the power of drifting, and of its improvisational powers, led logically to skating and parkour, whose genealogies I then shared. These practices are primarily, but not exclusively, engaged in by young men, who are often constituted as zeroes (unless and until they become [bankable] heroes such as skater Dan MacFarlane or *traceur* David Belle). Mindful of these varied constitutive narratives, I sought to understand how skating and tracing have been understood in burgeoning scholarly literature, examining their differences as disciplines, as practices, and in terms of their geographies, mobilities, and rhythms. Returning to moving images, I examined *Skateboarding Explained* and *Once is Never*, as well as Edwardes' (2009) treatise on parkour, and gestured to Castiglione's (2012) creative interventions in design for spaces of flow in our cities and settlements. Such works, and my analysis of them, suggested to me the existence of another kind of interval: a space or spaces in which we could be more generous in how we plan, develop, build, and live in our settlements. In this regard, near the end of the chapter, I sought insights from Goldberg, Leydon, and Scotto (2012), who write of the correlation between happiness, location, and connection, and then emphasized the point that to be moved on is a signature form of rejection, and one too often directed at young people. My suggestion that the spaces we create could be delightful *intergenerational* playgrounds may not be novel, but it does stress the importance of what Woodyer (2012) has eloquently described as a sense of play that could inform and underpin ethical generosity—which, I think, stands as an important supplement to established understandings of pleasure and of flourishing, and the modes of conduct we engage in to secure them.

In moving to share my encounters with and on the Circle Line, I also moved to a consideration of adult life, and the daily 'grind' of life in the labor force, itself characterized for so many of us by some form of regular, patterned, and spatially complex commuting. Noting that Brack may have captured on canvas a moment of pointed alienation and anomie on the part of office workers at 5pm in Melbourne's Collins Street, I posited that one nevertheless could imagine many of them returning home and to places in which they could dwell. I contrasted Brack's rush hour with another—Belle's playfully rhythmic two-minute trace from an office to an apartment via the rooftops of the city, which constitutes commuting as a form of play and challenges the logic of the urban grid and economistic understandings of transport. This work led me back to Lefebvre (2004), to thinking about what a rhythmanalyst might do, and to my own regular and deliberately auto-ethnographic experiences on the London Circle Line—events in which I experiment with his methods, with others' mobile methods, with haptic geographies, and play. Such practice has prompted an appreciation of London's status as an originary epicenter of commuting, and a desire to better understand the ways in which commuting is stressful and inconvenient, or may open up all kinds of interstices for dwelling-in-motion, as Sheller and Urry (2006) would have it. Such practice has also engendered in me a sense of anxiety that my conduct might first be seen, and then seen as suspicious, and that led me to reconceptualize the geographies, mobilities, and rhythms of commuting in terms of violent disruptions to it—not least the attacks on London Transport of 7/7. On these bases, I sought to engage with the literature on commuting—its history from the bipedal to the automobile, its gender differentials, and economistic and other modes of analyzing it, and the forms of conduct that have come to govern it. I then considered the relationship between commuting and multilocal living, gave thought to the numbers of commuters who are engaged in lengthy and complex trips, and reflected on Edensor's (2010) idea that dwelling-in-motion enables commuters to form and reform contingent, ephemeral, and affirmative relations (which Casey [1996] might cite as instantiations of a place as an event). Thereafter, I drew on Jensen's (2009) challenge to the binary thinking implicit in ideas about sedentarism and nomadism, and his conceptualization of commuting as a kind of armature. For Jensen, the armature allows us to move in different ways and to do 'other things'; perhaps this is to steal time, in Vannini's (2011) terms. Such practices are possible on many forms of transport, rail not the least among them, and the London Underground not the least among the great rail systems in the world. Thus, I outlined the genesis of that system which, while astounding, was not without significant diseconomies, negative effects, and critics—both contemporaneous (such as Mayhew) and contemporary (such as Cresswell). I made reference to Pearson (1859), a looming figure of reformist and utopian zeal who presages the desires that Wright (2010) enunciates in calling for real utopias now (albeit in a different political register and context). It was Pearson (1859) who understood the correlations between rail and land, land and housing, and housing and thriving; whose vision survived the fierce competition between rival operatives of the different lines—a vision which was, in some measure, enhanced by the colorful interventions of the American entrepreneur, Charles Tyson Yerkes, as well as the efforts of Albert Stanley and Frank Pick, whose working relationship forged the modern passenger experience. And then there is the sudden burst of disrupted daydreams, exemplified by the capacity of violence to reach into the city's armatures in order to do maximum harm, not simply by dint of force, but by constituting each of us as a becoming bomb, in Packer's (2006) terms. Such acts, I argued, dismantle trust by radically ramping up insecurity, while simultaneously enabling the dangerous conflation of patriotism, the sovereign, sovereignty, and the idea of territory as 'a warning off'. Thus it was that I found myself at the end of the line in company, I thought, with Hoffman (2002), Elden (2010), and Massumi (2007), whose works remind us of the scale, lethality, and spectacle of terrorism, and of the cul-de-sac formed by ongoing debates about prevention, deterrence, and preemption.

I do not wish to trivialize what I have conveyed immediately above in relation to acts of terror. Nevertheless, in thinking about the ways in which adults age (and more of us do, as life expectancy increases globally), it now strikes me that we also constitute the narratives of aging as another kind of undeclared war, and aging has induced many forms of anxiety, and many kinds of preventive, deterring, or preemptive conduct, not least because of a fundamental fear of dying. Thus, in chapter six, which is the second of three focused primarily on adults aged from around forty to around seventy, my task was to examine and unsettle how we constitute aging. I considered the ways in which aging has been seen as an inevitable and progressive onset of decline and incompetence, a view exemplifying just one form of ageism among many. I took account of such narratives of aging and of counternarratives that constitute the passage of one's decades as a mature adult in terms of their being an interval of long duration in which it might be possible to flourish. It was interesting to me that (in both scholarly terms and personal ones) I play in the conceptual and practical space that is 'adult fitness', and think it critically important to value and honor the body, its geographical range and reach, capacity to move, and astounding rhythms. In this light, I drew on Lefebvre's (2004) ideas of dressage and remarked on the utility of holding ourselves in particular bearing, engaging in repetitive acts to be disciplined—to follow particular ways of being. In so doing, I recognized the need to approach the narratives of fitness and aging critically, for there are all manner of biopolitical traps here. Allied to that recognition, I acknowledged that fitness is highly context dependent, and described the ways in which different gender identifications, different social contexts, and socio-economic groupings give effect to our understandings of fitness, aging, and their geographies, mobilities, and rhythms. I noted the ways in which septuagenarian Ernestine Shepherd exemplifies some of those, tracing how her fitness regime implicates a range of scales from the circuitry of cells to international body-sculpting circuits. I also commented on Shepherd's identification with ideas of salvation, desirability, discipline, and creative modes of habituation. I now see her journey, at least in part, as a kind of living memorialization to her sister Velvet—an insight I could not have made without, perhaps, thinking about Miriam Hyman and the loss that her sister Esther felt at her death on 7/7. And so I observed that if we are—in a sense—lucky enough, we will bear witness to our aging. Because it has become clear that so many more of us are likely to do so, there has been growing interest in arresting decline, fighting age, engaging—in Biggs's (1997) terms—in a masquerade of fluid identity choices, defying processes of senescence in ways that can be skin-deep or can invite us to deepen our practices and learn to move with skill, self-care, and a sense of *joie de vivre*. Hence, the idea that life begins at . . . [some indeterminate age that gets further and further from where we are now] and, as a corollary of that, the development of narratives of positive aging, which also require careful and critical responses such as represented by Nordbakke and Schwanen (2013), or Phoenix and Grant (2009), or Phoenix and Smith (2011), or McCormack (1999). His ideas about the geographies of fitness being woven from shifting landscapes of risk remain deeply discerning and map onto O'Malley's (2000) engagement with the constitution of the prudential citizen consumer, dispositions of risk, and regimes of practice. And then, I came to a point where I was able to argue that, on balance, there are indeed manifold benefits to moving it—to remaining rhythmically and geographically engaged in the world: dividends such as self-regulation, self-efficacy, and the alignment of expectations and outcomes; functional and structural resilience (including by developing better muscle mass and psychomotor performances, and better cardiovascular fitness); balance; and the reduced incidence of disease.

Notwithstanding the foregoing, we all enter the borderlands of the undiscovered country. It has been interesting to me that part of the impetus to commit to writing this book was my experience of encountering *Cradle to Grave*. Although my ardent desire is that we thrive through the life-course by creating—and aiding others to create—the spaces in which to flourish, I have found need to explicitly acknowledge in this work that we also die through the life-course. Aristotle might have been correct: there may be the possibility that we flourish somehow even when dead; and it is important to me that our descendants flourish as a result of the conduct of our lives. Either way, that we die is both an everyday reality and an extraordinary one, as Maddrell (2013) emphasizes. Thus, without being saccharine or sentimental, I wanted to examine the journey to the undiscovered country by drawing on examples that touch the work that I have presented here. I began with Overington's (2012) account of baby Leo and those who came together to celebrate his life, in part for the intrinsic good of such action, and in part to create the space in which they might themselves flourish again. I described how children worldwide are so vulnerable to preventable diseases and other symptoms of immiseration, structural disadvantage, and gross inequities, and noted that both they and climate change are intensifiers for risk, as Strauss (2012) puts it. My argument, clearly, was that we need to remain vigilant and pay more attention to their care; in doing so, we might make astonishing inroads to help optimize their journeys. Moving to delineate what typifies causes of death globally, I pointed out that a significant number of people die as a result of road trauma. A small number of them, each number a life, are skaters—at least some of whom were engaged in the hybrid geographies, mobilities, and rhythms of playful commuting when they were killed. Assuredly, I know that better design might have prevented this; and that there is room and cause for more generosity in the provision of space. Different forms of commuting and violent disruptions to it concerned me thereafter, and I noted the manner in which the deaths of commuters as a result of acts of terror compelled others to remember and provide living memorials to those who were killed; and although I cannot in any way countenance what was done, it did not escape me that Khan described himself and his co-conspirators as somehow fundamentally distressed by the death and dying of distant others, elsewhere, who arguably were themselves the victims of spiraling cycles of preemption. In then thinking about death in mid-life, I referred to the demise of fitness advocate Jim Fixx (1977)—who died running. I noted the passing of Christopher Hitchens (2012), whose experience of dying from esophageal cancer included, on the one hand, receiving letters from harsh fundamentalist critics who confused his illness with laryngeal cancer, and celebrated the fact that he, a confirmed atheist, had lost his voice and freedom to speak, and, on the other hand, the salve of friendship, which he most valued. I wrote of Derek Miller (2011), who blogged his own death, and determined that when it was done, it was done, and in so doing had the last word.

So, having acknowledged that people die through the life-course, and having noted that their passing has particular geographies, mobilities, and rhythms, and varied implications for how we conduct ourselves and might flourish, I turned my attention to the oldest-old, and to the idea that we wear out, that our bodies have limits, are finite, and decompose. I pondered the fact that, if cognizant, we are there for our dying but not for our death, which raised for me questions about the biopower of medicine and its capacity to render the body at least partly legible; and about the advent of the status of the *faux vivant* (which brings into view the specter of the pursuit of eternity), and the rise of *homo sacer* (and puzzles about what is worth living and what can be allowed to die). These reflections led me to consider tools such as advance care directives, and a widespread impulse to die in our homes despite the fact that many of us will journey into formal institutions of aged and then palliative care. These journeys and places engender other geographies, mobilities, and rhythms, and ask us to consider the difference between quietude and acquiescence. Here, the notion of *déprise* is useful: a gentle reorganization of daily life without any insinuation of a downward spiral; an idea that invites us to rethink what we mean by frailty and see it as part of a virtuous and complete life that enables us to let go and, by doing so, flourish in our final days. As part of that letting go among the oldest-old, and among the rest of us, may be processes to face certain fears—perhaps of dying alone, of suffering, or of uncertainty about what is after death; processes to face what dying means to our next of kin and to caregivers; processes that acknowledge the effects of *our* dying on *their* bodies, lives, geographies, mobilities, and rhythms. This is, perhaps, what Froggatt and her colleagues (2010) meant when they wrote of the living-dying interval—that ongoing need to hold open spaces of hope and a praxis of care around death and dying; or what Mann (2012) was referring to when she observed that there is need to think about how living with those who need our care as they die enriches our lives; or what Leavy (2011) and Cowley (2013) thought when, in different ways, they underscored the importance to us of being able to write the last chapter well.

Now in these latter days of writing, I have found myself turning to other women's engagements with Aristotelian thought, wanting to bring to this work the benefit of their perspectives. Cheshire Calhoun (2004), for example, reminds us to celebrate the idea that people exist also as *homo reparans*—moral agents, caring and repairing creatures. She asks us to challenge the tendency in moral philosophy to view "moral persons as fully formed adults who are not located in inegalitarian societies, who do not experience long periods of dependency on others, whose moral life is not hedged with contingencies, whose pleasures and passions are not constitutive elements of their moral life, and for whom reason giving is unaffected by the narratives we tell of our lives or by dynamics of social interaction" (6). Her invitation resonates deeply with me, for I see great merit in care and repair: these are labors of love, if you like, that are possible across vast spatial and temporal scales, and which salve and sooth and enrich. Calhoun (2004) also emphasizes the point that moral and political philosophy are not sharply bounded pursuits. Rather, there is a need to labor at their edges with each other (and with other disciplines), unsettling attempts to neatly circumscribe them and accounting for subjects traditionally marginalized from the debate, not least are those in caregiving roles. Again, my intention has been to highlight such matters. In turn, Marcia Homiak (2004, 25) points to the importance in Aristotelian thought of *praxeis*—activities complete in themselves, "at each moment of their being", and the separation of *praxeis* (and philosophical contemplation as its highest expression) from *kinēseis* or *poiēseis* (production, doing, 'mere' movement). Homiak notes Aristotle's view that "the best life is the life on unimpeded, continuous activity . . . But ordinary lives, he thinks, fall short of the best life because most human beings cannot fully realize their abilities to think and to know" (24). She challenges this view, arguing that "reasonably unimpeded activity is within most people's grasp and that most people, once they experience unimpeded activity, will be attracted to virtue" (24). A virtuous life of the kind she envisages—a caring, thoughtful, and hopeful life—is one that will engender flourishing. To this estimation, I would add that such a life will also account for the ways in which we are placed in the world in vital conjunctures (as Johnson-Hanks would have it) and as spatial, mobile, and rhythmic creatures. Indeed, as Jane Bennett (2010, 13) has surmised, "Each human is a heterogeneous compound of wonderfully vibrant, dangerously vibrant, matter. If matter itself is lively, then not only is the difference between subjects and objects minimized, but the status of the shared materiality of all things is elevated . . . The ethical aim becomes to distribute value more generously".

Works Cited

AARP. 2011. *Ernestine Shepherd: Body Builder—My Generation*. YouTube. [http://www.youtube.com/watch?v=4wXFSczN6Rw] accessed on 6 January 2014.

Abrahamsson, S. and P. Simpson. 2011. "The limits of the body: boundaries, capacities, thresholds." *Social & Cultural Geography* 12 (4): 331–338.

Aday, R.H., G.C. Kehoe, and L.A. Farney. 2006. "Impact of senior center friendships on aging women who live alone." *Journal of Women & Aging* 18 (1): 57–73.

Adey, P. 2010a. *Aerial Life: Spaces, Mobilities, Affects*. Chichester: Wiley-Blackwell.

Adey, P. 2010b. *Mobility*. London and New York: Routledge.

Adey, P., D. Bissell, K. Hannam, P. Merriman, and M. Sheller (Eds). 2014. *The Routledge Handbook of Mobilities*. Abingdon and New York: Routledge.

Adey, P., D. Bissell, D. McCormack, and P. Merriman. 2012. "Profiling the passenger: mobilities, identities, embodiments." *Cultural Geographies* 19 (2): 169–193.

Adger, W.N., J. Barnett, K. Brown, N. Marshall, and K. O'Brien. 2012. "Cultural dimensions of climate change impacts and adaptation." *Nature Climate Change* 3 (2): 112–117.

Agamben, G. 1998. *Homo Sacer: Sovereign Power and Bare Life*. (Trans. D. Heller-Roazen). Stanford: Stanford University Press

Agarwal, S.K. 2012. "Cardiovascular benefits of exercise." *International Journal of General Medicine* 5 (June): 541–545.

Aitken, S. 2001. *The Geographies of Young People: the Morally Contested Spaces of Identity*. London: Routledge.

Alaimo, S. 2010. "Bits of life: Feminism at the intersections of media, bioscience, and technology (review)." *Configurations* 18 (3): 477–480.

Albert, S.M., A. Im, and V.H. Raveis. 2002. "Public health and the second 50 years of life." *American Journal of Public Health* 92 (8): 1214–1216.

Allen, M.J. and S.D. Brown. 2011. "Embodiment and living memorials: The affective labour of remembering the 2005 London bombings." *Memory Studies* 4 (3): 312–327.

Allen, M.J. and A. Bryan. 2011. "Editorial: Remembering the 2005 London bombings: Media, memory, commemoration." *Memory Studies* 4 (3): 263–268.

Allen, W. 1973. *Everything You Always Wanted to Know About Sex But Were Afraid to Ask*. United Artists.

Aly, A. and L. Green. 2010. "Fear, anxiety and the state of terror." *Studies in Conflict & Terrorism* 33 (3): 268–281.

Anderson, J. 2004. "Talking whilst walking: a geographical archaeology of knowledge." *Area* 36 (3): 254–261.

Anderson, K. and S.J. Smith. 2001. "Editorial. Emotional geographies." *Transactions of the Institute of British Geographers* NS 26 (1): 7–10.

Andersson, M., A.K. Ekwall, I.R. Hallberg, and A.K. Edberg. 2010. "The experience of being next of kin to an older person in the last phase of life." *Palliative and Supportive Care* 8 (1): 17–26.

Andersson, M., I.R. Hallberg, and A.K. Edberg. 2008. "Old people receiving municipal care, their experiences of what constitutes a good life in the last phase of life: a qualitative study." *International Journal of Nursing Studies* 45 (6): 818–828.

Andrews, G.J., E. Hall, B. Evans, and R. Colls. 2012. "Moving beyond walkability: On the potential of health geography." *Social Science & Medicine* 75 (11): 1925–1932.

Andrews, G.J., R. Kearns, P. Kontos, and V. Wilson. 2006 "Their finest hour: older people, oral histories, and the historical geography of social life." *Social & Cultural Geography* 7 (2): 153–177.

Angel, J. 2009. *Once Is Never: Training with Parkour Generations*. London: Parkour Generations [1 hour 56 minutes].

Angel, J. 2011. *Ciné Parkour: A Cinematic and Theoretical Contribution to the Understanding of the Practice of Parkour*. London: Brunel University Screen Media Research Centre, School of Arts, Doctor of Philosophy.

Angel, J.L., J.K. Montez, and R.J. Angel. 2011. "A window of vulnerability: Health insurance coverage among women 55 to 64 years of age." *Women's Health Issues* 21 (1): 6–11.

Annas, G.J. 2007. "The Supreme Court and abortion rights." *New England Journal of Medicine* 356 (21): 2201–2207.

Anonymous. 2012. "The most beautiful minds in America". *Huffington Post*. Posted April 23. [http://www.huffingtonpost.com/2012/04/23/the-most-beautiful-minds-_n_1445265.html] accessed on 13 January 2014.

Ansell, N. 2009. "Childhood and the politics of scale: descaling children's geographies?" *Progress in Human Geography* 33 (2): 190–209.

Anton, A.L. 2012. "Respecting one's elders: in search of an ontological explanation for the asymmetry between the proper treatment of dependent adults and children." *Philosophical Papers* 41 (3): 397–419.

Antoninetti, M. and M. Garrett. 2012. "Body capital and the geography of aging." *Area* 44 (3): 364–370.

Appadurai, A. 1996. *Modernity at Large: Cultural Dimensions of Globalization*. Minneapolis and London: University of Minnesota Press.

Archambault, J. 2012. " 'It can be good there too': home and continuity in refugee children's narratives of settlement." *Children's Geographies* 10 (1): 35–48.

Aristotle. 350 BCE(a). *De Anima*. (Trans. J.A. Smith). Boston: MIT Classics.

Aristotle. 350 BCE(b). *The Nicomachean Ethics*. (Trans. W.D. Ross). Boston: MIT Classics.

Ashford, D. 2013. *London Underground: A Cultural Geography*. Liverpool: Liverpool University Press.

Atkinson, M. 2009. "Parkour, anarcho-environmentalism, and poiesis." *Journal of Sport & Social Issues* 33 (2): 169–194.

Augé, M. 1995. *Non-Places. Introduction to an Anthropology of Supermodernity*. (Trans. J. Howe). London and New York: Verso.

Ault, D. 1974. *Visionary Physics: Blake's Response to Newton*. Chicago: University of Chicago Press.

Australian Bureau of Statistics. 2007. 4829.0.55.001—Health of children in Australia: A snapshot, 2004–05. [http://www.abs.gov.au/ausstats/abs@.nsf/mf/4829.0.55.001/] accessed on 16 March 2014.

Australian Bureau of Statistics. 2010. 4704.0—The health and welfare of Australia's Aboriginal and Torres Strait islander peoples. [http://www.abs.gov.au/AUSSTATS/abs@.nsf/lookup/4704.0Chapter218Oct+2010] accessed on 4 October 2011.

Australian Bureau of Statistics. 2012. 3303.0 Causes of Death, Australia, Table 13.1 Fetal, neonatal and perinatal deaths, 2001–2010, Table 13.4 Perinatal deaths, main condition in the mother, by sex, 2006–2010. [http://www.abs.gov.au/AUSSTATS/abs@.nsf/DetailsPage/3303.02010?OpenDocument] accessed on 21 March 2014.

Australian Bureau of Statistics. 2013. 3303.0 Causes of Death, Australia, 2011. Table 1.3 Underlying cause of death, selected causes by age of death, numbers and rates, Australia, 2011. [http://www.abs.gov.au/AUSSTATS/abs@.nsf/Details Page/3303.02011?OpenDocument] accessed on 22 March 2014.

Australian Government. Australian Institute of Family Studies. 2013. Age of consent laws. [http://www.aifs.gov.au/cfca/pubs/factsheets/a142090/] accessed on 16 July 2013.

Australian Government. Australian Taxation Office. 2013. Individual tax return instructions 2013. Car and travel expenses 2013. [http://www.ato.gov.au/Individuals/Tax-return/2013/Tax-return/Deduction-questions-D1-D10/Car-and-travel-expenses/] accessed on 9 November 2013.

Australian Government. Department of Human Services. 2013. Youth Allowance. [http://www.humanservices.gov.au/customer/services/centrelink/youth-allo wance] accessed on 15 September 2013.

Australian Government. Department of Infrastructure and Regional Development. Bureau of Infrastructure, Transport and Regional Economics. 2013. Road deaths Australia. [http://www.bitre.gov.au/publications/ongoing/rda/files/RDA_Dec13. pdf] accessed on 17 Maqrch 2014.

Australian Health Society. 1900. *Twenty-Five Years' Record of the Work and Progress of the Society, with List of Subscribers, Library Catalogue, Society's Publications, &c.* Melbourne: Australian Health Society.

Australian Psychological Society. 2014. "Ageing positively: what is positive ageing?" [http://www.psychology.org.au/publications/tip_sheets/ageing/] accessed on 9 January 2014.

Aziz, N.M., J.L. Miller, and J.R. Curtis. 2012. "Palliative and end-of-life care research: Embracing new opportunities." *Nursing Outlook* 60 (6): 384–390.

Badrinarayana, D. 2010a. "Global warming: a second coming for international law?" *Washington Law Review* 85 (2): 253–294.

Badrinarayana, D. 2010b. "International law in a time of climate change, sovereignty loss, and economic unity." *Proceedings of the Annual Meeting—American Society of International Law* 104: 256–259.

Baer, P. 2010. "Adaptation to climate change." In S.M. Gardiner, S. Caney, D. Jamieson, and H. Shue (Eds), *Climate Ethics. Essential Readings*, 247–262. Oxford: Oxford University Press.

Baker, L. and E. Gringart. 2009. "Body image and self-esteem in older adulthood." *Ageing and Society* 29 (6): 977–995.

Barnett, J. 2012. "On the risks of engineering mobility to reduce vulnerability to climate change: insights from a small island state." In K. Hastrup and K.F. Olwig (Eds), *Climate Change and Human Mobility. Global Challenges to the Social Sciences*, 169–189. Cambridge: Cambridge University Press.

Barnett, J. and J. Campbell. 2010. *Climate Change and Small Island States: Power, Knowledge and the South Pacific.* London and Washington, D.C.: Earthscan.

Bartlett, S. 2008. "Climate change and urban children: Impacts and implications for adaptation in low-and middle-income countries." *Environment & Urbanization* 20 (2): 501–519.

Bassoli, A., J. Brewer, K. Martin, P. Dourish, and S. Mainwaring. 2007. "Underground aesthetics: Rethinking urban computing." *IEEE Pervasive Computing* 6 (3): 39–45.

Bateman Novaes, S. and T. Salem. 1998. "Embedding the embryo." In J. Harris and S. Holm (Eds), *The Future of Human Reproduction: Ethics, Choice, and Regulation*, 100–126. Oxford: Clarendon Press.

Beauchamp, M.R., A.V. Carron, S. McCutcheon, and O. Harper. 2007. "Older adults' preferences for exercising alone versus in groups: considering contextual congruence." *Annals of Behavioral Medicine* 33 (2): 200–206.

Bell, N. 2008. "Ethics in child research: rights, reason and responsibilities." *Children's Geographies* 6 (1): 7–20.

Bennett, J. 2010. *Vibrant Matter: A Political Ecology of Things*. Durham and London: Duke University Press.

Bergmann, S. 2008. "The beauty of speed or the discovery of slowness—why do we need to rethink mobility?". In S. Bergmann and T. Sager (Eds), *The Ethics of Mobilities. Rethinking Place, Exclusion, Freedom and Environment*, 13–24. Aldershot: Ashgate.

Bergmann, S. and T. Sager. 2008. "In between standstill and hypermobility: introductory remarks to a broader discourse." In S. Bergmann and T. Sager (Eds), *The Ethics of Mobilities: Rethinking Place, Exclusion, Freedom and Environment*, 1–12. Aldershot: Ashgate.

Bergson, H. 1946 [1923]. *The Creative Mind: An Introduction to Metaphysics*. (Trans. M.L. Andison). New York: Philosophical Library.

Bertram, G. 2006. "Introduction: The MIRAB model in the twenty-first century." *Asia Pacific Viewpoint* 47 (1): 1–13.

Betzold, C., P. Castro, and F. Weiler. 2011. *AOSIS in the UNFCCC Negotiations: from Unity to Fragmentation?* Zurich: Center for Comparative and International Studies (CIS), ETH-UZH.

Bherer, L., K.I. Erickson, and T. Liu-Ambrose. 2013. "A review of the effects of physical activity and exercise on cognitive and brain functions in older adults." *Journal of Aging Research* 2013 (Article ID 657508): 8pp.

Biggs, S. 1997. "Choosing not to be old? Masks, bodies and identity management in later life." *Ageing and Society* 17 (5): 553–570.

Bird, M.-L., K.D. Hill, and J.W. Fell. 2012. "A randomized controlled study investigating static and dynamic balance in older adults after training with Pilates." *Archives of Physical Medicine and Rehabilitation* 93 (1): 43–49.

Birk, T. 2012. "Relocation of reef and atoll island communities as an adaptation to climate change: learning from experience in Solomon Islands." In K. Hastrup and K.F. Olwig (Eds), *Climate Change and Human Mobility. Global Challenges to the Social Sciences*, 81–109. Cambridge: Cambridge University Press.

Bissell, D. 2007. "Animating suspension: waiting for mobilities." *Mobilities* 2 (2): 277–298.

Bissell, D. 2009a. "Moving with others: the sociality of the railway journey." In P. Vannini (Ed.), *The Cultures of Alternative Mobilities: Routes Less Travelled*, 55–70. Aldershot: Ashgate.

Bissell, D. 2009b. "Travelling vulnerabilities: mobile timespaces of quiescence." *Cultural Geographies* 16 (4): 427–445.

Bissell, D. 2010. "Vibrating materialities: mobility–body–technology relations." *Area* 42 (4): 479–486.

Bissell, D. 2014a. "Encountering stressed bodies: Slow creep transformations and tipping points of commuting mobilities." *Geoforum* 51 (1): 191–201.

Bissell, D. 2014b. "Habits." In P. Adey, D. Bissell, K. Hannam, P. Merriman, and M. Sheller (Eds), *The Routledge International Handbook of Mobilities*, 483–492. London and New York: Routledge.

Blake, K. 2009. "The sea-dream: *Peter Pan* and *Treasure Island*." *Children's Literature* 6 (1): 165–181.

Blunt, A. 1994. *Travel, Gender, and Imperialism: Mary Kingsley and West Africa*. New York: The Guilford Press.

Blunt, A. 2007. "Cultural geographies of migration: mobility, transnationality and diaspora." *Progress in Human Geography* 31 (5): 684–694.

Böhm, S., C. Jones, C. Land, and M. Paterson. 2006. "Part One Conceptualizing Automobility: Introduction: Impossibilities of automobility." *The Sociological Review* 54: 1–16.

Bond, C.J. and K. Lowton. 2011. "Geriatricians' views of advance decisions and their use in clinical care in England: qualitative study." *Age and Ageing* 40 (4): 450–456.

Bongso, A. and M. Richards. 2004. "History and perspective of stem cell research." *Best Practice & Research Clinical Obstetrics & Gynaecology* 18 (6): 827–842.

Bonham, J. 2006. "Part Two. Governing automobility. Transport: disciplining the body that travels." *The Sociological Review* 54: 55–74.

Borden, I. 2001. *Skateboarding, Space and the City: Architecture and the Body.* Oxford and New York: Berg Publishers.

Botman, D.P.J. and M. Kumar. 2008. Social Science Research Network Working Paper Series: "Global Aging Pressures: Impact of Fiscal Adjustment, Policy Cooperation, and Structural Reforms." 3 April. [http://ssrn.com/abstract=1997237] accessed on 8 January 2014.

Bouchonville, M., R. Armamento-Villareal, K. Shah, N. Napoli, D.R. Sinacore, C. Qualls, and D.T. Villareal. 2013. "Weight loss, exercise or both and cardiometabolic risk factors in obese older adults: results of a randomized controlled trial." *International Journal of Obesity.* [http://www.nature.com/ijo/journal/vaop/ncurrent/full/ijo2013122a.html] accessed on 10 January 2014.

Breaugh, J.A. and A.M. Farabee. 2012. "Telecommuting and flexible work hours: alternative work arrangements that can improve the quality of work life." In J.A. Breaugh and A.M. Farabee (Eds), *Work and Quality of Life,* 251–274. Amsterdam: Springer.

Brickell, K. 2012. "'Mapping' and 'doing' critical geographies of home." *Progress in Human Geography* 36 (2): 225–244.

Brigham and Women's Hospital Boston. 2014. About the Positive Aging Resource Center. [http://www.positiveaging.org/] accessed on 9 January 2014.

British Broadcasting Corporation. 2005a. London bomber: Text in full, Thursday, 1 September. [http://news.bbc.co.uk/2/hi/uk_news/4206800.stm] accessed on 30 November 2013.

British Broadcasting Corporation. 2005b. Toll climbs in Egyptian attacks, Saturday, 23 July [http://news.bbc.co.uk/2/hi/middle_east/4709491.stm] accessed on 20 March 2014.

British Broadcasting Corporation. 2006. David Belle—Rush Hour. [http://www.youtube.com/watch?v=SAMAr8y-Vtw] accessed on 15 September 2013.

British Broadcasting Corporation. 2012. Middle age begins at 55 years, survey suggests, Monday, 17 September [http://www.bbc.co.uk/news/education-19622330] accessed on 28 December 2013.

British Broadcasting Corporation. no date. 7 July bombings. Overview [http://news.bbc.co.uk/2/shared/spl/hi/uk/05/london_blasts/what_happened/html/] accessed on 29 November 2013.

Brunner, C. 2011. "Nice-looking obstacles: parkour as urban practice of deterritorialization." *AI & Society* 26 (2): 143–152.

Buckley, B., A. Goulding, A. Newman, E. Pringle, and C. Whitehead. 2007. *Artists' Insights: A study of the practices, and the learning outcomes and impact, of visual artists working with young people and educators/facilitators in art gallery contexts.* Newcastle-upon-Tyne: International Centre for Cultural and Heritage Studies, Newcastle University.

Burges-Watson, D. and E. Stratford. 2008. "Feminizing risk at a distance: critical observations on the constitution of a preventive technology for HIV/AIDS." *Social & Cultural Geography* 9 (4): 353–371.

Burkett, M. 2011. "The Nation *Ex-Situ*: On climate change, deterritorialized nationhood and the post-climate era." *Climate Law* 2 (3): 345–374.

Burrell, N. 2012. *The Future is Your Choice: Art, Urbanism, and Activism in the (Post) industrial City. Senior Capstone Projects.* Paper 124. Poughkeepsie, New York: Vassar College.

Burton, D., J. Mustelin, and P. Urich. 2011. *Climate Change Impacts on Children in the Pacific. A Focus on Kiribati and Vanuatu. Advocacy Paper.* Paris: UNICEF.

Buttimer, A. 1993. *Geography and the Human Spirit.* Baltimore and London: The John Hopkins University Press.

Buttimer, A. and D. Seamon. 1980. *The Human Experience of Space and Place.* New York: St Martin's Press.

Bynner, J. and M. Wadsworth. 2011. "Generation and change in perspective." *A Companion to Life Course Studies. The Social and Historical Context of the British Birth Cohort Studies,* 203–221. London and New York: Routledge.

Calasanti, T. 2005. "Ageism, gravity, and gender: Experiences of aging bodies." *Generations* 29 (3): 8–12.

Calhoun, C. 2004. "Introduction." In C. Calhoun (Ed.), *Setting the Moral Compass: Essays by Women Philosophers,* 1–19. Oxford: Oxford University Press.

Caney, S. 2010. "Climate change, human rights, and moral thresholds." In S.M. Gardiner, S. Caney, D. Jamieson, and H. Shue (Eds), *Climate Ethics. Essential Readings,* 162–177. Oxford: Oxford University Press.

Caparros, M., L. Laski, and V. Bernhardtz. 2009. *At the Frontier: Young People and Climate Change. Youth Supplement to the UNFPA State of World Population Report 2009.* New York: UNFPA.

Carlander, I., E. Sahlberg Blom, I. Hellström, and B.M. Ternestedt. 2011. "The modified self: family caregivers' experiences of caring for a dying family member at home." *Journal of Clinical Nursing* 20 (7–8): 1097–1105.

Carr, J. 2010. "Legal geographies—skating around the edges of the law: urban skateboarding and the role of law in determining young peoples' place in the city." *Urban Geography* 31 (7): 988–1003.

Carr, J. 2012. "Activist research and city politics: Ethical lessons from youth-based public scholarship." *Action Research* 10 (1): 61–78.

Casey, E.S. 1996. "How to get from space to place in a fairly short time: phenomenological prolegomena." In S. Feld and K.H. Basso (Eds), *Senses of Place,* 13–52. Santa Fe: School of American Research Press.

Castells, M. 2010. *The Rise of the Network Society: The Information Age: Economy, Society, and Culture* (Vol. 1, 2nd Edition). Oxford: Wiley-Blackwell.

Castiglione, R. 2012. *Dead Space.* Mercy, Michigan: University of Detroit Mercy, School of Architecture, Masters of Architecture.

Castro, J.P., F.A. El-Atat, S.I. McFarlane, A. Aneja, and J.R. Sowers. 2003. "Cardiometabolic syndrome: pathophysiology and treatment." *Current Hypertension Reports* 5 (5): 393–401.

Chambers, A.F. and K.S. Chambers. 2007. "Five takes on climate and cultural change in Tuvalu." *The Contemporary Pacific* 19 (1): 294–306.

Child, L. 2007. What is street training? [http://www.streettraining.org/content/what-street-training] accessed on 14 September 2013.

Child, L. 2009. Manchester Street Training 10. [http://www.youtube.com/watch?v=HpHj8zaBhKk] accessed on 14 September 2013.

Child, L. 2010. "Street training in Loughborough: on climbing, testing, penetrating, playing with, nurturing, building and/or pissing on boundaries—physical, mental and social—in the perpetual making of public space." *Visual Studies* 25 (1): 85–86.

Chiu, C. 2009. "Contestation and conformity street and park skateboarding in New York City public space." *Space and Culture* 12 (1): 25–42.

Chow, B.D.V. 2010. "Parkour and the critique of ideology: Turn-vaulting the fortresses of the city." *Journal of Dance and Somatic Practices* 2 (2): 143–154.

Cibelli, J. and K. Wang. 2010. "Creating 'human' embryos. Interspecies somatic cell nuclear transfer: not yet 'healthy' human embryos." In J. Nisker, F. Baylis, I. Karpin, C. McLeod, and R. Mykitiuk (Eds), *The 'Healthy' Embryo. Social, Biomedical, Legal and Philosophical Perspectives*, 61–69. Cambridge: Cambridge University Press.

Clifford, J. 1997. *Routes: Travel and Translation in the Late Twentieth Century.* Boston: Harvard University Press.

Cloke, P. 2002. "Deliver us from evil? Prospects for living ethically and acting politically in human geography." *Progress in Human Geography* 26 (5): 587–604.

Collins, C.A. and A. Opie. 2010. "When places have agency: Roadside shrines as traumascapes." *Continuum: Journal of Media & Cultural Studies* 24 (1): 107–118.

Conley, O. 2012. Lennart Nilsson. Embryo Project Encyclopedia. [http://stsrepo sitory.mbl.edu/bitstream/handle/10776/1995/embryo127589.xhtml?sequence=1] accessed on 13 March 2013.

Connell, J. 2013. *Islands at Risk? Environments, Economies, and Contemporary Change.* Cheltenham: Edward Elgar.

Corner, G.W. 1954. *George Linius Streeter 1873–1948. A Biographical Memoir.* Washington, D.C.: National Academy of Sciences.

Cosgrove, D. 1994. "Contested global visions: one-world, whole-Earth, and the Apollo space photographs." *Annals of the Association of American Geographers* 84 (2): 270–294.

Cottingham, J. 2012. "The question of ageing." *Philosophical Papers* 41 (3): 371–396.

Coupland, J. 2009. "Time, the body and the reversibility of ageing: commodifying the decade." *Ageing and Society* 29 (6): 953–976.

Cowley, C. 2013. "The last chapter in the story: a place for Aristotle's eudaemonia in the lives of the terminally ill." *Online Journal of Health Ethics* 3 (1): Article 5, 1–10.

Crampton, J.W. 2011. "Cartographic calculations of territory." *Progress in Human Geography* 35 (1): 92–103.

Crampton, J.W. and S. Elden. 2007. *Space, Knowledge and Power: Foucault and Geography.* Aldershot: Ashgate.

Cresswell, T. 2006. *On the Move. Mobility in the Modern Western World.* London and New York: Routledge.

Cresswell, T. 2010a. "Mobilities I: Catching up." *Progress in Human Geography* 35 (4): 550–558.

Cresswell, T. 2010b. "Towards a politics of mobility." *Environment and Planning D: Society and Space* 28 (1): 17–31.

Cresswell, T. and D. Dixon. 2002. *Engaging Film: Geographies of Mobility and Identity.* Lanham, Maryland: Rowman & Littlefield Publishing Incorporated.

Cresswell, T. and P. Merriman. 2011. *Geographies of Mobilities: Practices, Spaces, Subjects.* Aldershot: Ashgate.

Cronin, A.M. 2008. "Mobility and market research: outdoor advertising and the commercial ontology of the city." *Mobilities* 3 (1): 95–115.

Daily Mail Reporter. 2011. " 'Here it is. I am dead': Canadian blogger announces own death in heart-rending post from beyond the grave." *UK Daily Mail*, 6 May. [http://www.dailymail.co.uk/news/article-1384031/Derek-K-Miller-announces-death-blog-post-grave.html] accessed on 2 February 2014.

Daniels, M. 2012. Generation Yamakazi (API Productions, Majestic Force, France 2, SBS-TV Australie). [http://www.youtube.com/watch?v=fOYpHLHg6io] accessed on 16 October 2013.

Darbyshire, P. 2010. "Heroines, hookers and harridans: Exploring popular images and representations of nurses and nursing." In J. Daly, S. Speedy, and D. Jackson (Eds), *Contexts of Nursing*, 51–63. Sydney: Churchill Livingston/Elsevier.

Daskalaki, M., A. Stara, and M. Imas. 2008. "The 'Parkour Organisation': inhabitation of corporate spaces." *Culture and Organization* 14 (1): 49–64.

Davies, D.J. and C.W. Park. 2012. *Emotion, Identity, and Death: Mortality Across Disciplines. Introduction*. Aldershot: Ashgate Publishing.

Davies, R. 2005. "Mothers' stories of loss: their need to be with their dying child and their child's body after death." *Journal of Child Health Care* 9 (4): 288–300.

de Certeau, M. 1984. *The Practice of Everyday Life*. (Trans. S. Rendell). Berkeley: University of California Press.

de Freitas, E. 2011. "Parkour and the build environment: Spatial practices and the plasticity of school buildings." *Journal of Curriculum Theorizing* 27 (3): 209–220.

De Sousa, P. 2010. "Parthenogenesis and other strategies to create human embryos for stem cell research and regenerative medicine." In J. Nisker, F. Baylis, I. Karpin, C. McLeod, and R. Mykitiuk (Eds), *The 'Healthy' Embryo. Social, Biomedical, Legal and Philosophical Perspectives*, 70–83. Cambridge: Cambridge University Press.

Dean, M. 1999. *Governmentality: Power and Rule in Modern Society*. Thousand Oaks, London, New Delhi: Sage Publications.

Dean, M. 2002. "Powers of life and death beyond governmentality." *Cultural Values* 6 (1–2): 119–138.

Debord, G. 1956. "Theory of the *dérive*" (translated by Ken Knabb). *Les Lèvres Nues* #9. Reprinted in *Internationale Situationniste* #2 (December 1958). [http://www.cddc.vt.edu/sionline/si/theory.html] accessed on 25 September 2013.

Debord, G. 1967. *The Society of the Spectacle*. (Trans. D. Nicholson-Smith). New York: Zone Books.

Deleuze, G. and Guattari, F. 1988. *A Thousand Plateaus: Capitalism and Schizophrenia*. London: Athlone.

DeLoughrey, E. 2007. *Routes and Roots: Navigating Caribbean and Pacific Island Literatures*. Honolulu: University of Hawai'i Press.

DeLyser, D. and D. Sui. 2013. "Crossing the qualitative-quantitative divide II. Inventive approaches to big data, mobile methods, and rhythmanalysis." *Progress in Human Geography* 37 (2): 293–305.

Denfeld, D. 1974. "Woody and Rag are dead: A consideration of the three stages of American automobile consciousness." *The Journal of Popular Culture* 8 (1): 148–154.

Derrida, J. 2005. *Sovereignties in Question. The Poetics of Paul Celan*. (T. Dutoit and O. Pasenen Eds.). New York: Fordham University Press.

Dinces, S. 2011. " 'Flexible opposition': Skateboarding subcultures under the rubric of late capitalism." *The International Journal of the History of Sport* 28 (11): 1512–1535.

Dishman, R.K. 1994. "Motivating older adults to exercise." *Southern Medical Journal* 87 (5): S79–S82.

Dolan, P., T. Peasgood and M. White. 2008. "Do we really know what makes us happy? A review of the economic literature on the factors associated with subjective well-being." *Journal of Economic Psychology* 29 (1): 94–122.

Doughty, K. 2011. "Walking and Well-being: Landscape, Affect, Rhythm." Southhampton: University of Southhampton, Department of Geography and Environment, Doctor of Philosophy.

Dryden, L., S. Arata, and E. Massie. 2010. "Robert Louis Stevenson and popular culture." *Nordic Journal of English Studies* 9 (3): 11–24.

Dubow, S. 2012. *Ourselves Unborn: A History of the Fetus in Modern America.* New York: Oxford University Press.

Duffy, M., G. Waitt, A. Gorman-Murray, and C. Gibson. 2011. "Bodily rhythms: corporeal capacities to engage with festival spaces." *Emotion, Space and Society* 4 (1): 17–24.

Dumas, A. and S. Laberge. 2005. "Social class and ageing bodies: Understanding physical activity in later life." *Social Theory & Health* 3 (3): 183–205.

Dumas, A. and S. Laforest. 2009. "Skateparks as a health-resource: Are they as dangerous as they look?" *Leisure Studies* 28 (1): 19–34.

Eastell, R. 2013. "Identification and management of osteoporosis in older adults." *Medicine* 41 (1): 47–52.

Edensor, T. 2010a. "Commuter: Mobility, rhythm and commuting." In T. Cresswell and P. Merriman (Eds), *Geographies of Mobilities: Practices, Spaces, Subjects*, 189–203. Aldershot: Ashgate.

Edensor, T. 2010b. "Introduction: Thinking about rhythm and space." In T. Edensor (Ed.), *Geographies of Rhythm: Nature, Place, Mobilities and Bodies*, 1–18. Aldershot: Ashgate.

Edensor, T. 2010c. "Walking in rhythms: Place, regulation, style and the flow of experience." *Visual Studies* 25 (1): 69–79.

Edensor, T. 2014. "Rhythm and arrhythmia." In P. Adey, D. Bissell, K. Hannam, P. Merriman, and M. Sheller (Eds), *The Routledge Handbook of Mobilities*, 163–171. Abingdon and New York: Routledge.

Edwardes, D. 2009. *The Parkour and Freerunning Handbook.* London: Harper Collins.

Edwardes, D. 2010. Parkour History, Parkour Generations. [http://www.parkourgene rations.com/article/parkour-history] accessed on 10 October 2013.

Elden, S. 2004a. "Rhythmanalysis: An introduction." In H. Lefebvre, *Rhythmanalysis: Space, Time and Everyday Life.* London and New York: Continuum, vii–xv.

Elden, S. 2004b. *Understanding Henri Lefebvre.* London and New York: Continuum.

Elden, S. 2005. "Missing the point: Globalization, deterritorialization and the space of the world." *Transactions of the Institute of British Geographers NS* 30 (1): 8–19.

Elden, S. 2009. *Terror and Territory. The Spatial Extent of Sovereignty.* Minneapolis and London: University of Minnesota Press.

Elden, S. and Crampton, J.W. 2007. Introduction. Space, Knowledge and Power: Foucault and Geography. In J.W. Crampton and S. Elden (Eds), *Space, Knowledge and Power: Foucault and Geography.* Farnham: Ashgate Publishing Limited, 1–16.

Ellis, C., T.E. Adams, and A.P. Bochner. 2011. "Autoethnography: an overview." Forum: Qualitative Social Research, 12 (1) Article 10. [http://www. qualitative-research.net/index.php/fqs/article/viewArticle/1589/3095%20 on%2001%20July%202011] accessed on 6 January 2014.

Ellis, C. and A.P. Bochner. 2000. "Autoethnography, personal narrative, reflexivity." In N.K. Denzin and Y.S. Lincoln (Eds), *Handbook of Qualitative Research, Second Edition*, 733–768. Thousand Oaks, London, New Delhi: Sage Publications.

Eman, J. 2012. "The role of sports in making sense of the process of growing old." *Journal of Aging Studies* 26 (4): 467–475.

England, M.A. 1983. *A Colour Atlas of Life Before Birth: Normal Fetal Development.* London: Wolfe Medical Publications Ltd.

Ermarth, E.D. 1992. *Sequel to History: Postmodernism and the Crisis of Representational Time.* Princeton: Princeton University Press.

Ermarth, E.D. 2011. *History in the Discursive Condition.* London and New York: Routledge.

Evans, B. 2008. "Geographies of youth/young people." *Geography Compass* 2 (5): 1659–1680.

Evans, G.W. and P. Kim. 2010. "Multiple risk exposure as a potential explanatory mechanism for the socioeconomic status–health gradient." *Annals of the New York Academy of Sciences* 1186 (1): 174–189.

Fahlman, M., A. Morgan, N. McNevin, R. Topp, and D. Boardley. 2007. "Combination training and resistance training as effective interventions to improve functioning in elders." *Journal of Aging and Physical Activity* 15 (2): 195–205.

Farbotko, C. 2010a. " 'The global warming clock is ticking so see these places while you can': Voyeuristic tourism and model environmental citizens on Tuvalu's disappearing islands." *Singapore Journal of Tropical Geography* 31 (2): 224–238.

Farbotko, C. 2010b. "Wishful sinking: Disappearing islands, climate refugees and cosmopolitan experimentation." *Asia Pacific Viewpoint* 51 (1): 47–60.

Farbotko, C. 2012. "Skilful seafarers, oceanic drifters or climate refugees? Pacific people, news value and the climate refugee crisis." In K. Moore, B. Gross, and T. Threadgold (Eds), *Migration and the Media*, 119–142. New York: Peter Lang Publishing.

Fekete, M. 2006. *Strength Training for Seniors. How to Rewind Your Biological Clock.* Toronto: Key Porter Books.

Felstiner, J. 2001. *Paul Celan: Poet, Survivor, Jew.* New Haven CT and London: Yale University Press.

Ferrari, L., M. Berlingerio, F. Calabrese and J. Reades. 2014 in press. "Improving the accessibility of urban transportation networks for people with disabilities." *Transportation Research Part C: Emerging Technologies.* [http://www.science direct.com/science/article/pii/S0968090X13002209] accessed on 3 March 2014.

Finin, G.A. 2002. "Small is viable: The global ebbs and flows of a Pacific atoll nation." *Pacific Island Development Series, East West Center Working Papers* 15 (April): 1–29.

Fixx, J. 1977. *The Complete Book of Running.* New York: Random House.

Flourish Over 50. 2014. "Flourish over 50. Midlife reinventions to experience life to the fullest! Feature: How not to look old." [http://www.flourishover50.com/tag/how-not-to-look-old/] accessed on 2 March 2014.

Ford, M.F. (Ed.). 1905. *The Soul of London: A Survey of a Modern City.* London: Alston Rivers.

Foucault, M. 1972. *The Archeology of Knowledge and the Discourse on Language.* (Trans. A. Sheridan). New York: Pantheon Books.

Foucault, M. 1973. *The Birth of the Clinic.* (Trans. A. Sheridan). London: Tavistock.

Foucault, M. 1975. *Discipline and Punish: the Birth of the Prison.* (Trans. A. Sheridan). London: Peregrine.

Foucault, M. 1976. *The History of Sexuality, Volume 1: An Introduction.* (Trans. R. Hurley). Melbourne: Penguin.

Foucault, M. 1980. "Questions on geography: an interview with the editors of the journal Herodote." In C. Gordon (Ed.), *Power/Knowledge: Selected Interviews and Other Writings 1972–1977*, 63–77. Brighton: Harvester Press.

Foucault, M. 1982. "The subject and power." *Critical Inquiry* 8 (4): 777–795. [Pages 777–785 written in English by Foucault; the balance of pages 785–795 were translated from the French by Lesley Sawyer].

Foucault, M. 1986. "Of other spaces." *Diacritics* 16 (1): 22–27.

Foucault, M. 1991. "Governmentality." In G. Burchell, C. Gordon, and P. Milller (Eds), *The Foucault Effect: Studies in Governmentality*, 87–104. London: Harvester Wheatsheaf.

Foucault, M. 2007. *Security, Territory, Population.* New York: Picador.

Foucault, M. 2008. *The Birth of Biopolitics.* New York: Picador.

Founten, L. 2013. "Adelaide woman allowed to try to have dead husband's baby." *ABC News,* Thursday, 19 December. [http://www.abc.net.au/news/2013-12-19/

adelaide-woman-allowed-to-try-to-have-dead-husband27s-baby/5167814]
accessed on 2 February 2014.

Franklin, R.E. and R. Gosling. 1953. "Molecular configuration in sodium thymo-nucleate." *Nature* 171 (4356): 740–741.

Franklin, S. 2006. "The cyborg embryo: our path to transbiology." *Theory, Culture & Society* 23 (7–8): 167–187.

Froggatt, K., J. Hockley, and D. Parker. 2010. "A system lifeworld perspective on dying in long term care settings for older people: Contested states in contested places." *Health & Place* 17 (1): 263–268.

Fry, P.S. 1990. "A factor analytic investigation of home-bound elderly individuals' concerns about death and dying, and their coping responses." *Journal of Clinical Psychology* 46 (6): 737–748.

Gardiner, S.M. 2010a. "Introductory overview." In S.M. Gardiner, S. Caney, D. Jamieson, and H. Shue (Eds), *Climate Ethics. Essential Readings*, 3–35. Oxford: Oxford University Press.

Gardiner, S.M. 2010b. "A perfect moral storm. Climate change, intergenerational ethics, and the problem of corruption." In S.M. Gardiner, S. Caney, D. Jamieson, and H. Shue (Eds), *Climate Ethics. Essential Readings*, 87–98. Oxford: Oxford University Press.

Gehl Architects. 2004. *Towards a Fine City for People. Public Spaces and Public Life—London 2004*. London and Copenhagen: Gehl Architects, Central London Partnerships, Transport for London.

Gehl, J. 2011. *Life Between Buildings: Using Public Space*. Washington, D.C.: Island Press.

German Institute for Economic Research (DIW Berlin Deutsches Institut für Wirtschaftsforschung e.V.). 2013. G-SOEP—German Socio-Economic Panel (DIW Berlin) [http://www.eui.eu/Research/Library/ResearchGuides/Economics/Statistics/DataPortal/GSOEP.aspx] accessed on 16 November 2013.

Ghaye, T. 2010. "Editorial: In what ways can reflective practices enhance human flourishing?" *Reflective Practice: International and Multidisciplinary Perspectives* 11 (1): 1–7.

Giele, J.Z. and G.H. Elder. 1998. "Life course research. Development of a field." In J.Z. Giele and G.H. Elder (Eds), *Methods of Life Course Research: Qualitative and Quantitative Approaches*, 5–27. Thousand Oaks, London, New Delhi: Sage Publications.

Gilleard, C. and P. Higgs. 2011. "Ageing abjection and embodiment in the fourth age." *Journal of Aging Studies* 25 (2): 135–142.

Goggins, R. 2012. "Aristotle on happiness and death." *Southwest Philosophy Review* 28 (1): 63–71.

Goldberg, A., K.M. Leyden, and T.J. Scotto. 2012. "Untangling what makes cities liveable: happiness in five cities." *Urban Design and Planning* 165 (3): 127–136.

Goldenberg, M. and W. Shooter. 2009. "Skateboard park participation: A means-end analysis." *Journal of Youth Development* 4 (4): 12pp.

Golombek, S.B. 2006. "Children as citizens." *Journal of Community Practice* 14 (1–2): 11–30.

Goonewardena, K. 2011. "Henri Lefebvre." In G. Ritzer and J. Stepnisky (Eds), *The Wiley-Blackwell Companion to Major Social Theorists. Volume 2: Contemporary Social Theorists*, 44–64. Chichester: Wiley-Blackwell.

Gordon, G.M. and M.M. Cuttic. 1994. "Exercise and the aging foot." *Southern Medical Journal* 87 (5): S36–S41.

Gorman-Murray, A. 2008. "Masculinity and the home: a critical review and conceptual framework." *Australian Geographer* 39 (3): 367–379.

Gosden, R. and B. Lee. 2010. "Portrait of an oocyte: our obscure origin." *The Journal of Clinical Investigation* 120 (4): 973–983.

Government of Canada. Transport Canada. 2010. Canadian Motor Vehicle Traffic Collision Statistics: 2010 [http://www.tc.gc.ca/eng/motorvehiclesafety/tp-1317.htm] accessed on 17 March 2014.

Government of Canada. National Seniors Council. 2011. Report of the National Seniors Council on Volunteering Among Seniors and Positive and Active Aging. [http://www.seniorscouncil.gc.ca/eng/research_publications/volunteering/page06.shtml] accessed on 9 January 2014.

Graham, A. and R. Fitzgerald. 2010. "Progressing children's participation: Exploring the potential of a dialogical turn." *Childhood* 17 (3): 343–359.

Graves, J.E., M.L. Pollock, and J.F. Carroll. 1994. "Exercise, age, and skeletal muscle function." *Southern Medical Journal* 87 (5): S17–S22.

Gregor, K. 2013. "The rise of the 'extreme commuter.'" *BBC News Online.* [http://www.bbc.com/news/magazine-25498136] accessed on 29 March 2014.

Grenier, A. 2007. "Constructions of frailty in the English language, care practice and the lived experience." *Ageing and Society* 27 (3): 425–446.

Gross, J. 1984. "James F. Fixx dies jogging; author on running was 52." Obituaries. *New York Times,* July 22. [http://www.nytimes.com/1984/07/22/obituaries/james-f-fixx-dies-jogging-author-on-running-was-52.html] accessed on 16 January 2014.

Guerlac, S. 2006. *Thinking in Time: an Introduction to Henri Bergson.* Ithaca: Cornell University Press.

Gurmankin, A.D., D. Sisti, and A.L. Caplan. 2004. "Embryo disposal practices in IVF clinics in the United States." *Politics and the Life Sciences* 22 (2): 4–8.

Guss, N. 2011. "Parkour and the multitude: Politics of a dangerous art." *French Cultural Studies* 22 (1): 73–85.

Hägerstrand, T. 1968. *Innovation Diffusion as a Spatial Process.* Chicago: University of Chicago Press.

Hall, C., P. Thomson, and L. Russell. 2007. "Teaching like an artist: The pedagogic identities and practices of artists in schools." *British Journal of Sociology of Education* 28 (5): 605–619.

Hannam, K., M. Sheller, and J. Urry. 2006. "Editorial: Mobilities, immobilities and moorings." *Mobilities* 1 (1): 1–22.

Hanson, S. 2010. "Gender and mobility: New approaches for informing sustainability." *Gender, Place & Culture* 17 (1): 5–23.

Haraway, D. 1991. *Simians, Cyborgs, and Women: the Reinvention of Nature.* New York: Routledge.

Haraway, D. 1997. "The virtual speculum in the new world order." *Feminist Review* 55 (Spring): 22–72.

Hart, R.A. 1997. *Children's Participation: the Theory and Practice of Involving Young Citizens in Community Development and Environmental Care.* Abingdon: Earthscan.

Harvard Medical School. 2006. "Our balancing act." *Harvard Health Letter* 31 (10): 1–3.

Hastrup, K. and K.F. Olwig. 2012. "Introduction: climate change and human mobility." In K. Hastrup and K.F. Olwig (Eds), *Climate Change and Human Mobility. Global Challenges to the Social Sciences,* 1–20. Cambridge: Cambridge University Press.

Hein, J.R., J. Evans, and P. Jones. 2008. "Mobile methodologies: Theory, technology and practice." *Geography Compass* 2 (5): 1266–1285.

Heiskanen, B. 2012. *The Urban Geography of Boxing: Race, Class, and Gender in the Ring. Routledge Research in Sport, Culture and Society, Volume 13.* London and New York: Routledge.

Held, V. 1989. "Birth and death." *Ethics* 99 (2): 362–388.

Herod, A. 2010. *Scale*. Abingdon and New York: Routledge.

Higgs, P., M. Leontowitsch, F. Stevenson, and I.R. Jones. 2009. "Not just old and sick—the 'will to health' in later life." *Ageing and Society* 29 (5): 687–707.

Hilti, N. 2009. "Here, there and in-between: on the interplay of multilocal living, space, and inequality." In T. Ohnmacht, H. Maksim, and M.M. Bergman (Eds), *Mobilities and Inequality*, 145–164. Aldershot: Ashgate Publishing.

Hitchens, C. 2012. *Mortality*. New York and Boston: Twelve. Hachette Book Group.

Hoffman, B. 2002. "Rethinking terrorism and counterterrorism since 9/11." *Studies in Conflict & Terrorism* 25 (5): 303–316.

Hogan, C.L., J. Mata, and L.L. Carstensen. 2013. "Exercise holds immediate benefits for affect and cognition in younger and older adults." *Psychology and Aging* 28 (2): 587–594.

Homiak, M. 2004. "Virtue and the skills of ordinary life." In C. Calhoun (Ed.), *Setting the Moral Compass: Essays by Women Philosophers*, 23–42. Oxford: Oxford University Press.

Hopkins, P. and N. Bell. 2008. "Interdisciplinary perspectives: ethical issues and child research." *Children's Geographies* 6 (1): 1–6.

Hopkins, P. and M. Hill. 2008. "Pre-flight experiences and migration stories: The accounts of unaccompanied asylum-seeking children." *Children's Geographies* 6 (3): 257–268.

Horton, J., P. Kraftl, and F. Tucker. 2008. "The challenges of 'Children's Geographies': A reaffirmation." *Children's Geographies* 6 (4): 335–348.

Horton, R. 2012. The Global Burden of Disease Study 2010, Published December 13, 2012. [http://www.thelancet.com/themed/global-burden-of-disease] accessed on 28 December 2013.

Howell, O. 2008. "Skatepark as neoliberal playground: Urban governance, recreation space, and the cultivation of personal responsibility." *Space and Culture* 11 (4): 475–496.

Howitt, R. 2002. "Scale and the other: Levinas and geography." *Geoforum* 33 (3): 299–303.

Hubbard, P. 2008. "Here, there, everywhere: The ubiquitous geographies of heteronormativity." *Geography Compass* 2 (3): 640–658.

Hubbard, P. and K. Lilley. 2004. "Pacemaking the modern city: The urban politics of speed and slowness." *Environment and Planning D: Society and Space* 22 (2): 273–294.

Hulme, M. 2009. *Why We Disagree About Climate Change. Understanding Controversy, Inaction and Opportunity*. Cambridge: Cambridge University Press.

Hurd Clarke, L. and A. Korotchenko. 2011. "Aging and the body: A review." *Canadian Journal on Aging/La Revue canadienne du vieillissement* 1 (1): 1–16.

Illich, I. 1976. *Medical Nemesis: The Expropriation of Health*. New York: Pantheon.

Ingold, T. 2004. "Culture on the ground: The world perceived through the feet." *Journal of Material Culture* 9 (3): 315–340.

Ingold, T. 2005. "The eye of the storm: Visual perception and the weather." *Visual Studies* 20 (2): 97–104.

Ingold, T. 2010. "Footprints through the weather world: Walking, breathing, knowing." *Journal of the Royal Anthropological Institute* 16 (1 [Supplement]): S121–S139.

International Monetary Fund. 2011. Australia—World Economic Outlook Database. [http://www.imf.org/external/pubs/ft/weo/2011/02/weodata/weorept.aspx?pr.x=75&pr.y=17&sy=2009&ey=2016&scsm=1&ssd=1&sort=country&ds=.&br=1&c=193&s=NGDPDPC&grp=0&a=] accessed on 10 January 2012.

Ionesco, E. 1960. *Rhinoceros and Other Plays*. (Trans. D. Prouse). New York: Grove Press.

Jain, S. 1998. "Mysterious delicacies and ambiguous agents: Lennart Nilsson in National Geographic." *Configurations* 6 (3): 373–394.

James, A. 1986. "Learning to belong: The boundaries of adolescence." In A.P. Cohen (Ed.), *Symbolising Boundaries: Identity and Diversity in British Cultures*, 155–171. Manchester: Manchester University Press.

James, S. 1990. "Is there a 'place' for children in geography?" *Area* 22 (3): 278–283.

Jamieson, D. 1992. "Ethics, public policy, and global warming." *Science, Technology & Human Values* 17 (2): 139–153.

Jans, M. 2004. "Children as citizens: Towards a contemporary notion of child participation." *Childhood* 11 (1): 27–44.

Jeffrey, C. 2010. "Geographies of children and youth I: Eroding maps of life." *Progress in Human Geography* 34 (4): 496–505.

Jeffrey, C. 2012. "Geographies of children and youth II: Global youth agency." *Progress in Human Geography* 36 (2): 245–253.

Jeffrey, C. 2013. "Geographies of children and youth III: Alchemists of the revolution?" *Progress in Human Geography* 37 (1): 145–152.

Jenkins, M. 2013. "Parkour classes are helping pensioners stay agile and active." *The Guardian*, Wednesday, 28 August. [http://www.theguardian.com/society/2013/aug/28/parkour-classes-pensioners-agile-active] accessed on 11 October 2013.

Jensen, O.B. 2009. "Flows of meaning, cultures of movements—Urban mobility as meaningful everyday life practice." *Mobilities* 4 (1): 139–158.

Jenson, A., J. Swords, and M. Jeffries. 2012. "The accidental youth club: Skateboarding in Newcastle-Gateshead." *Journal of Urban Design* 17 (3): 371–388.

Johnson Hanks, J. 2002. "On the limits of life stages in ethnography: Toward a theory of vital conjunctures." *American Anthropologist* 104 (3): 865–880.

Johnston, L. 1996. "Flexing femininity: Female body-builders refiguring 'the body'." *Gender, Place & Culture* 3 (3): 327–340.

Jones, O. 2012. *Chavs: The Demonization of the Working Class*. London: Verso Books.

Jones, P. 2012. "Sensory indiscipline and affect: A study of commuter cycling." *Social & Cultural Geography* 13 (6): 645–658.

Kaiser, H.J. 2009. "Mobility in old age." *Journal of Applied Gerontology* 28 (4): 411–418.

Kaplan, C. 1998. *Questions of Travel: Postmodern Discourses of Displacement*. Durham: Duke University Press.

Kasbarian, J.A. 1996. "Mapping Edward Said: Geography, identity, and the politics of location." *Environment and Planning D: Society and Space* 14 (5): 529–557.

Katz, C. 2001. "Vagabond capitalism and the necessity of social reproduction." *Antipode* 33 (4): 709–728.

Katz, C. 2004. *Growing Up Global: Economic Restructuring and Children's Everyday Lives*. Minneapolis: University of Minnesota Press.

Katz, S. 2000. "Busy bodies: Activity, aging, and the management of everyday life." *Journal of Aging Studies* 14 (2): 135–152.

Kearns, R. and D. Collins. 2010. "Health geography." In T. Brown, S. McLafferty, and G. Moon (Eds), *A Companion to Health and Medical Geography*, 15–32. Chichester: Blackwell Publishing Ltd.

Kearns, R. and G. Moon. 2002. "From medical to health geography: Novelty, place and theory after a decade of change." *Progress in Human Geography* 26 (5): 605–625.

Keeley III, C.J. 2006. "Subway searches: Which exception to the warrant and probable cause requirements applies to suspicionless searches of mass transit passengers to prevent terrorism?" *Fordham Law Review* 74 (6): 3231–3295.

Kelley, J. 2011. "Climate change and small island states: Adrift in a rising sea of legal uncertainty." *Sustainable Development Law and Policy* 11 (2): 56–57.

Kendrick, Z.V., S. Nelson-Steen, and K. Scafidi. 1994. "Exercise, aging, and nutrition." *Southern Medical Journal* 87 (5): S50–S60.
Kennedy, C. and C. Lukinbeal. 1997. "Towards a holistic approach to geographic research on film." *Progress in Human Geography* 21 (1): 33–50.
Kenner, A.M. 2008. "Securing the elderly body: Dementia, surveillance, and the politics of aging in place." *Surveillance & Society* 5 (3): 252–269.
Kesby, M. 2007. "Methodological insights on and from Children's Geographies." *Children's Geographies* 5 (3): 193–205.
Khan, C.A. 2009. "Go play in traffic: Skating, gender, and urban context." *Qualitative Inquiry* 15 (6): 1084–1102.
Kidder, J.L. 2012. "Parkour, the affective appropriation of urban space, and the real/virtual dialectic." *City & Community* 11 (3): 229–253.
Knox, P.L. 2005. "Creating ordinary places: Slow cities in a fast world." *Journal of Urban Design* 10 (1): 1–11.
Kofman, E. and E. Lebas. 1996. "Part I. Introduction. Lost in transposition—time, space and the city." In H. Lefebvre, *Writings on Cities*. Oxford: Blackwell Publishers, 3–62.
Kullman, K. and C. Palludan. 2011. "Rhythmanalytical sketches: agencies, school journeys, temporalities." *Children's Geographies* 9 (3–4): 347–359.
Latour, B. 2011. "Visualisation and cognition: Drawing things together." *Knowledge and Society* 6 1–40.
Law, J. 1994. *Organizing Modernity*. Oxford: Blackwell Publishers.
Law, R. 1999. "Beyond 'women and transport': Towards new geographies of gender and daily mobility." *Progress in Human Geography* 23 (4): 567–588.
Lawler, J., with, M. Urbano, M. Deasey, N. Kusumawardani, J. Pardosi, D. Burton, J. Mustelin, P. Urich, P. Nenova-Knight, R. Perez, A.Y. Loyzaga, G.T. Narisma, C. Vincente, D. Olaguer, and M. Patel. 2011. *Children's Vulnerability to Climate Change and Disaster Impacts in East Asia and the Pacific*. Bangkok: UNICEF.
Leavy, S.A. 2011. "The last of life: Psychological reflections on old age and death." *The Psychoanalytic Quarterly* 80 699–715.
Leckie, S. 2008. "Climate change and displacement: Human rights implications." *Forced Migration Review* 31 18–19.
Lee, N. and J. Motzkau. 2011. "Navigating the bio-politics of childhood." *Childhood* 18 (1): 7–19.
Lefebvre, H. 1991. *The Production of Space*. (Trans. D. Nicholson-Smith). Oxford Blackwell Publishers.
Lefebvre, H. 1996. *Writings on Cities*. (E. Kofman and E. Lebas Eds.). Oxford: Blackwell Publishers.
Lefebvre, H. 2004. *Rhythmanalysis: Space, Time, and Everyday Life*. (S. Elden and G. Moore, Trans. S. Elden Ed.). London and New York: Continuum.
Lefebvre, H. and C. Régulier. 2004. "The Rhythmanalytical Project." Originally published as "Le projet rythmanalytique." *Communications* 41." In H. Lefebvre, *Rhythmanalysis. Space, Time and Everyday Life*. London and New York: Continuum, 73–83.
Levinson, D. 2008. "Density and dispersion: The co-development of land use and rail in London." *Journal of Economic Geography* 8 (1): 55–77.
Limacher, M.C. 1994. "Aging and cardiac function: Influence of exercise." *Southern Medical Journal* 87 (5): S13–S16.
Lloyd, D., D. Wilson, F. Tuddenham, G. Goodman, and A. Bhagat. 2013. Reported Road Casualties. Great Britain. 2012. [https://www.gov.uk/government/uploads/system/uploads/attachment_data/file/269601/rrcgb-2012-complete.pdf] accessed on 17 March 2014.
Longhurst, R. 2006. "A pornography of birth: Crossing moral boundaries." *ACME: An International E-journal for Critical Geographies* 5 (2): 209–229.

Lord, S., C. Després, and T. Ramadier. 2011. "When mobility makes sense: A qualitative and longitudinal study of the daily mobility of the elderly." *Journal of Environmental Psychology* 31 (1): 52–61.

Lorimer, H. 2011. "Walking: new forms and spaces for studies of pedestrianism." In T. Cresswell and P. Merriman (Eds), *Geographies of Mobilities: Practices, Spaces, Subjects*, 19–34. Farnham: Ashgate.

Low, N., H. Douglas, N. Kessler, J. Hurst, and E. Stratford. 2010. *Fresh! A Map of a Dream of the Future. Exploring Island Life, Climate Change, Resilience*. Hobart: School of Geography and Environmental Studies, University of Tasmania.

Lowenthal, D.T., D.A. Kirschner, N.T. Scarpace, M. Pollock, and J. Graves. 1994. "Effects of exercise on age and disease." *Southern Medical Journal* 87 (5): S5–S13.

Lupton, D. 1999. "Monsters in metal cocoons: 'Road rage' and cyborg bodies." *Body and Society* 5 (1): 57–72.

Lustenberger, T., P. Talving, G. Barmparas, B. Schnüriger, L. Lam, K. Inaba, and D. Demetriades. 2010. "Skateboard-related injuries: Not to be taken lightly. A national trauma databank analysis." *The Journal of Trauma and Acute Care Surgery* 69 (4): 924–927.

Lynn, J. and D.M. Adamson. 2003. *Living Well at the End of Life. Adapting Health Care to Serious Chronic Illness in Old Age*. Santa Monica: RAND Health.

Lyons, G., J. Jain, and D. Holley. 2007. "The use of travel time by rail passengers in Great Britain." *Transportation Research Part A: Policy and Practice* 41 (1): 107–120.

Lyons, G., J. Jain, Y. Susilo, and S. Atkins. 2013. "Comparing rail passengers' travel time use in Great Britain between 2004 and 2010." *Mobilities* 8 (4): 560–579.

Maayan, I. 2012. NOVA's *The Miracle of Life* (1983). Embryo Project Encyclopedia. http://stsrepository.mbl.edu/bitstream/handle/10776/2025/embryo127874.xhtml?sequence=1 accessed on 13 March 2013.

Macauley, D. 2000. "Walking the city: An essay on peripatetic practices and politics." *Capitalism Nature Socialism* 11 (4): 3–43.

MacFarlane, D. 2006. *Skateboarding Explained: The Instructional DVD*. United States: Mentality Skateboards, Skateboarder Magazine, Ollie Pops, and Lake Owen Camp.

MacKay, S. and C. Dallaire. 2013. "Skirtboarder net-a-narratives: Young women creating their own skateboarding (re)presentations." *International Review for the Sociology of Sport* 48 (2): 171–195.

Maddrell, A. 2013. "Living with the deceased: Absence, presence and absence-presence." *Cultural Geographies* 20 (4): 501–522.

Maddrell, A. and J. Sidaway (Eds). 2010. *Deathscapes. Spaces for Death, Dying, Mourning and Remembrance*. Aldershot: Ashgate.

Maienschein, J. and J.S. Robert. 2010. "What is an embryo and how do we know?". In J. Nisker, F. Baylis, I. Karpin, C. McLeod, and R. Mykitiuk (Eds), *The 'Healthy' Embryo. Social, Biomedical, Legal and Philosophical Perspectives*, 1–15. Cambridge: Cambridge University Press.

Malbon, B. 2002. *Clubbing: Dancing, Ecstasy, Vitality*. London and New York: Routledge.

Malone, K. 2012. "'The future lies in our hands': Children as researchers and environmental change agents in designing a child-friendly neighbourhood." *Local Environment* 18 (3): 372–395.

Malone, K. and C. Hartung. 2010. "Challenges of participatory practice with children." In B. Percy-Smith and N. Thomas (Eds), *A Handbook of Children and Young People's Participation: Perspectives from Theory and Practice*, 24–38. London and New York: Routledge.

Malpas, J. 2012. "Putting space in place: Philosophical topography and relational geography." *Environment and Planning D: Society and Space* 30 (2): 226–242.

Mann, H.S. 2012. "Ancient virtues, contemporary practices: An Aristotelian approach to embodied care." *Political Theory* 40 (2): 194–221.

Marmot, M. 2005. "Social determinants of health inequalities." *The Lancet* 365 (9464): 1099–1104.

Marston, S.A. 2000. "The social construction of scale." *Progress in Human Geography* 24 (2): 219–242.

Martin, D. 1953. *Sway [Quién será]*. Music by Luis Demetrio and English lyrics by Norman Gimbel.

Martin, T. and S. Vice. 2012. "Introduction." *Philosophical Papers* 41 (3): 331–333.

Massey, D. 1998. "The spatial construction of youth cultures." In G. Valentine, T. Skelton and D. Chambers (Eds), *Cool Places. Geographies of Youth Cultures*, 121–129. London and New York: Routledge.

Massumi, B. 2007. "Potential politics and the primacy of preemption." *Theory & Event* 10 (2): 1.

Matthews, H. 2003. "Inaugural editorial: Coming of age for children's geographies." *Children's Geographies* 1 (1): 3–5.

Mauss, M. 1973. "Techniques of the body." *Economy and Society* 2 (1): 70–88.

Mayer, T. (Ed.). 1999. *Gender Ironies of Nationalism: Sexing the Nation.* London and New York: Routledge.

McAdam, J. 2010. *Refusing 'Refuge' in the Pacific:(De) Constructing Climate-Induced Displacement in International Law. University of New South Wales Faculty of Law Research Series 2010. Working Paper 27.* Sydney: University of New South Wales Faculty of Law.

McAdam, J. and B. Saul. 2010. *An insecure climate for human security? Climate-induced displacement and international law. University of New South Wales Faculty of Law Research Series. Working Paper 59.* Sydney: University of New South Wales Faculty of Law.

McCormack, D. 1999. "Body shopping: Reconfiguring geographies of fitness." *Gender, Place & Culture* 6 (2): 155–177.

mCenter. The Mobilities Research and Policy Center. 2011. About mCentre. [http://mcenterdrexel.wordpress.com/about/] accessed on 4 October 2011.

McLeod, C. and F. Baylis. 2010. "Donating fresh versus frozen embryos to stem cell research: In whose interests?". In J. Nisker, F. Baylis, I. Karpin, C. McLeod, and R. Mykitiuk (Eds), *The 'Healthy' Embryo. Social, Biomedical, Legal and Philosophical Perspectives*, 171–186. Cambridge: Cambridge University Press.

McMichael, A.J. and E. Lindgren. 2011. "Climate change: Present and future risks to health, and necessary responses." *Journal of Internal Medicine* 270 (5): 401–413.

McNay, L. 1992. *Foucault and Feminism: Power, Gender and the Self.* Oxford: Polity Press.

McTavish, L. 2010. "A visual dialogue on 'healthy' human embryos from the sixteenth to the twenty-first centuries." In J. Nisker, F. Baylis, I. Karpin, C. McLeod, and R. Mykitiuk (Eds), *The 'Healthy' Embryo. Social, Biomedical, Legal and Philosophical Perspectives*, 97–115. Cambridge: Cambridge University Press.

Medline Plus. US National Library of Medicine and the National Institutes of Health. 2014. Death among children and adolescents. [http://www.nlm.nih.gov/medlineplus/ency/article/001915.htm] accessed on 16 March 2014.

Meier, L., L. Frers, and E. Sigvardsdotter. 2013. "Editorial. The importance of absence in the present: practices of remembrance and the contestation of absences." *Cultural Geographies* 20 (4): 423–430.

Mentality Skateboards. 2011. Skate with your mentality. [http://www.mentality skateboards.com/] accessed on 15 October 2013.

Merriman, P. 2013. "Rethinking mobile methods." *Mobilities* (DOI: 10.1080/17450101.2013.784540): 1–21.

Miciukiewicz, K. and G. Vigar. 2013. "Encounters in motion: Considerations of time and social justice in urban mobility research." In D. Henckel, S. Thomaier, B. Konecke, R. Zedda, and S. Stabilini (Eds), *Space-Time Design of the Public City*, 171–185. Amsterdam: Springer Link.

Middleton, J. 2010. "Sense and the city: Exploring the embodied geographies of urban walking." *Social & Cultural Geography* 11 (6): 575–596.

Millar, I. 2008. "It's the end of the world as we know it—human displacement, loss of States and climate change." *Amicus Curiae* 2008 (73): 1–2.

Miller, D.K. 2011. "The time will come." Derek K. Miller, Vancouver, Canada. [http://www.penmachine.com/2011/04/time-will-come] accessed on 16 January 2014.

Milligan, C. 2012. *There's No Place Like Home: Place and Care in an Ageing Society*. Aldershot: Ashgate.

Mimura, N., L. Nurse, R.F. McLean, J. Agard, L. Briguglio, P. Lefale, R. Payet, and G. Sem. 2007. "Small Islands". In M.L. Parry, O.F. Canziani, J.P. Palutikof, P.J. van der Linden, and C.E. Hanson (Eds), *Climate Change 2007: Impacts, Adaptation and Vulnerability. Contribution of Working Group II to the Fourth Assessment Report of the Intergovernmental Panel on Climate Change*, Chapter 16. Cambridge and New York: Cambridge University Press.

Miriam Hyman Memorial Trust. 2014. About the Miriam Hyman Memorial Trust. [http://www.miriam-hyman.com/] accessed on 20 March 2014.

Mitchell, J.K. 2003. "The fox and the hedgehog: Myopia about Homeland Security in U.S. policies on terrorism." *Research in Social Problems and Public Policy (Special Issue on Terrorism and Disaster: New Threats, New Ideas)* 11 53–72.

Mitova, V. 2012. "Age and agency." *Philosophical Papers* 41 (3): 335–369.

Monroe, B., D. Oliviere, and S. Payne. 2011. "Introduction: Social differences— the challenge for palliative care." In D. Oliviere, B. Monroe, and S. Payne (Eds), *Death, Dying, and Social Differences*, 1–7. Oxford: Oxford University Press.

Moore, A. 2008. "Rethinking scale as a geographical category: From analysis to practice." *Progress in Human Geography* 32 (2): 203–225.

Morgan, L.M. 2009. *Icons of Life: A Cultural History of Human Embryos*. Berkeley: University of California Press.

Morris, B. 2004. "What we talk about when we talk about 'walking in the city'." *Cultural Studies* 18 (5): 675–697.

Mortreux, C. and J. Barnett. 2009. "Climate change, migration and adaptation in Funafuti, Tuvalu." *Global Environmental Change* 19 (1): 105–112.

Mould, O. 2009. "Parkour, the city, the event." *Environment and Planning D: Society and Space* 27 (4): 738–750.

Mowl, G., R. Pain, and C. Talbot. 2000. "The ageing body and the homespace." *Area* 32 (2): 189–197.

msk1416. 2013. The best of the best of skateboarding. [http://www.youtube.com/watch?v=aK3zdJw40L8] accessed on 17 October 2013.

Mulvey, L. 1975/1991. "Visual pleasure and narrative cinema." In R.R. Warhol and D.P. Herndl (Eds), *Feminisms: An Anthology of Literary Theory and Criticism*, 438–448. New Jersey: Rutgers University Press.

Murray, L. 2009. "Looking at and looking back: Visualization in mobile research." *Qualitative Research* 9 (4): 469–488.

Nash, C. 1996. "Reclaiming vision: Looking at landscape and the body." *Gender, Place & Culture* 3 (2): 149–169.

Nash, C. 2005. "Geographies of relatedness." *Transactions of the Institute of British Geographers* 30 (4): 449–462.

National Galleries Scotland. nd. William McTaggart. *The Storm*. Caption and Biography. [http://www.nationalgalleries.org/collection/artists-a-] accessed on 26 June 2013.

National Gallery of Victoria. no date. Collection. John Brack. *Collins St. 5p.m. 1955*. [http://www.ngv.vic.gov.au/col/work/3161] accessed on 18 November 2013.

Neft, D. 1959. "Some aspects of rail commuting: New York, London, and Paris." *Geographical Review* 49 (2): 151–163.

Newman, K. 1996. *Fetal Positions: Individualism, Science, Visuality*. Stanford: Stanford University Press.

New Zealand Government. Ministry of Transport. 2014. Annual road toll historical information. [http://www.transport.govt.nz/research/roadtoll/annualroadtoll historicalinformation/] accessed on 17 March 2014.

Nicholson, C. and J. Hockley. 2011. "Death and dying in older people." In D. Oliviere, B. Monroe, and S. Payne (Eds), *Death, Dying, and Social Differences*, 101–109. Oxford: Oxford University Press.

Nicholson, S. 2011. *The 0 to 100 Project*. [http://www.0to100project.com/] accessed on 4 October 2011.

Nikander, P. 2009. "Doing change and continuity: Age identity and the micro-macro divide." *Ageing and Society* 29 (6): 863–881.

Nilsson, L. 1965. "Drama of life before birth." *Life* 58 (17): Cover. 35 x 27 cm.

Nilsson, L., B.G. Erikson, and C.O. Lofman. 1982/1986. *The Miracle of Life*. United States: Swedish Television Production in association with Boston WGBH Educational Foundation [first aired 15 February 1983].

Nordbakke, S. and T. Schwanen. 2013. "Well-being and mobility: A theoretical framework and literature review focusing on older people." *Mobilities* (DOI: 10.1080/17450101.2013.78542): 1–26.

Northcott, M.S. 2006. "In the waters of Babylon: The moral geography of the embryo." In C. Deane-Drummond and P.M. Scott (Eds), *Future Perfect? God, Medicine and Human Identity*, 73–86. London and New York: T&T Clark International. A Continuum Imprint.

NOVA. 1996. Interview with Lennart Nilsson. [http://www.pbs.org/wgbh/nova/odyssey/nilsson.html] accessed on 5 January 2013.

NOVA. 2013. About NOVA. [http://www.pbs.org/wgbh/nova/about/] accessed on 5 January 2013.

O'Brien, M. and B. Jack. 2010. "Barriers to dying at home: The impact of poor co-ordination of community service provision for patients with cancer." *Health & Social Care in the Community* 18 (4): 337–345.

O'Dell, T. 2009. "My soul for a seat: Commuting and the routines of mobility." In E. Shove, F. Trentmann and R. Wilk (Eds), *Time, Consumption and Everyday Life: Practice, Materiality and Culture*, 85–98. Oxford and New York: Berg Publishing.

O'Malley, P. 2000. "Uncertain subjects: Risks, liberalism and contract." *Economy and Society* 29 (4): 460–484.

Oakley, J.R. 1978. "Points from Letters. Skateboard injuries." *British Medical Journal* 1 (6105): 115.

Ohnmacht, T., H. Maksim, and M.M. Bergman. 2009. "Mobilities and inequality—an introduction." In T. Ohnmacht, H. Maksim, and M.M. Bergman (Eds), *Mobilities and Inequality*, 1–4. Aldershot: Ashgate.

Orehek, E., S. Fishman, M. Dechesne, B. Doosje, A.W. Kruglanski, A.P. Cole, B. Saddler, and T. Jackson. 2010. "Need for closure and the social response to terrorism." *Basic and Applied Social Psychology* 32 (4): 279–290.

Overington, C. 2012. "If this baby had been born fifteen years ago, these photos would not exist." *The Australian Women's Weekly* (October): 60–66.

Oxford University. 1971. [OED] *The Compact Edition of the Oxford English Dictionary.* Oxford: Oxford University Press.

Pachauri, R.K. and A. Reisinger. 2007. Intergovernmental Panel on Climate Change. Fourth Assessment Report: Climate Change 2007. Synthesis Report. Contribution of Working Groups I, II and III to the Fourth Assessment Report of the Intergovernmental Panel on Climate Change. Geneva: World Meteorological Organization and the United Nations Environment Program.

Packer, J. 2006. "Becoming bombs: Mobilizing mobility in the war of terror." *Cultural Studies* 20 (4–5): 378–399.

Packer, J. 2008. "Automobility and the driving force of warfare: From public safety to national security." In S. Bergmann and T. Sager (Eds), *The Ethics of Mobilities. Rethinking Place, Exclusion, Freedom and Environment,* 39–64. Aldershot: Ashgate.

Pain, R. 2003. "Youth, age and the representation of fear." *Capital & Class* 27 (2): 151–171.

Panelli, R., K. Nairn, and J. McCormack. 2002. " 'We make our own fun': Reading the politics of youth with(in) community." *Sociologia Ruralis* 42 (2): 106–130.

Parkour Conditioning. 2014. Parkour Injuries: 2012 Survey Results. [http://parkourconditioning.com/parkour-injuries-survey/] accessed on 21 March 2014.

Parkour Generations. 2013. Parkour Generations Services. [http://www.parkourgenerations.com/services/movies-and-stock-footage] accessed on 16 October 2013.

Pascal, C. and T. Bertram. 2009. "Listening to young citizens: The struggle to make real a participatory paradigm in research with young children." *European Early Childhood Education Research Journal* 17 (2): 249–262.

Paton, K. and P. Fairbairn-Dunlop. 2010. "Listening to local voices: Tuvaluans respond to climate change." *Local Environment* 15 (7): 687–698.

Paulson, S. 2005. "How various 'cultures of fitness' shape subjective experiences of growing older." *Ageing and Society* 25 (2): 229–244.

Paumgarten, N. 2007. "There and back again: The soul of the commuter. Annals of Transport" quoted in *The New Yorker,* April 16. [http://www.newyorker.com/reporting/2007/04/16/070416fa_fact_paumgarten?printable=true] accessed on 24 October 2013.

Pearson, C. 1859. *Proceedings at a Public Meeting Held at the London Tavern on the 1st December 1859. Central Railways in the Metropolis.* London: City of London.

Pfeifer, S. 2013. "Surfer calls on Roxy to change ads." *Los Angeles Times,* September 12. [http://articles.latimes.com/2013/sep/12/business/la-fi-roxy-ads-20130913] accessed on 14 October 2013.

Phoenix, C. 2010. "Auto-photography in aging studies: Exploring issues of identity construction in mature bodybuilders." *Journal of Aging Studies* 24 (3): 167–180.

Phoenix, C. and B. Grant. 2009. "Expanding the agenda for research on the physically active aging body." *Journal of Aging and Physical Activity* 17 (3): 362–379.

Phoenix, C. and B. Smith. 2011. "Telling a (good?) counterstory of aging: Natural bodybuilding meets the narrative of decline." *The Journals of Gerontology Series B: Psychological Sciences and Social Sciences* 66 (5): 628–639.

Pink, S. 2007. "Walking with video." *Visual Studies* 22 (3): 240–252.

Porter, G., J. Townsend, and K. Hampshire. 2012. "Children and young people as producers of knowledge." *Children's Geographies* 10 (2): 131–134.

Positive Ageing. no date. Welcome. Positive Ageing. [http://positiveageing.org.uk/] accessed on 9 January 2014.

Prout, A. 2005. *The Future of Childhood: Towards the Interdisciplinary Study of Children.* Abingdon and New York: Routledge.

Purcell, M. 2013. "To inhabit well: Counterhegemonic movements and the right to the city." *Urban Geography* 34 (4): 560–574.

Rao, R. 2010. "Property, privacy and other legal constructions of human embryos." In J. Nisker, F. Baylis, I. Karpin, C. McLeod, and R. Mykitiuk (Eds), *The 'Healthy' Embryo. Social, Biomedical, Legal and Philosophical Perspectives*, 32–44. Cambridge Cambridge University Press.

Rawlinson, C. and M. Guaralda. 2011. Play in the city: Parkour and architecture. In The First International Postgraduate Conference on Engineering, Designing and Developing the Built Environment for Sustainable Wellbeing. Brisbane: Queensland University of Technology.

Rayfuse, R. and E. Crawford. 2011. *Climate Change, Sovereignty and Statehood*. Sydney: University of Sydney.

Reading, A. 2011. "The London bombings: Mobile witnessing, mortal bodies and globital time." *Memory Studies* 4 (3): 298–311.

Reagan, L.J. 2012. "Ourselves unborn: A history of the fetus in modern America." *Journal of American History* 99 (3): 867–868.

Reich, O., A. Signorell, and A. Busato. 2013. "Place of death and health care utilization for people in the last 6 months of life in Switzerland: A retrospective analysis using administrative data." *BMC Health Services Research* 13 (1): 116–126.

Relph, E. 2000. *Place and Placelessness*. London: Pion.

Renz, M., M.S. Mao, D. Bueche, T. Cerny, and F. Strasser. 2013. "Dying is a transition." *American Journal of Hospice and Palliative Medicine* 30 (3): 283–290.

Robbins, M. 2004. Pearson, Charles. [http://www.oxforddnb.com/view/article/38367] accessed on 31 October 2013.

Roberts, J., R. Hodgson, and P. Dolan. 2011. " 'It's driving her mad': Gender differences in the effects of commuting on psychological health." *Journal of Health Economics* 30 (5): 1064–1076.

Rogoff, I. 2000. *Terra Infirma: Geography's Visual Culture*. London and New York: Routledge.

Romero-Arenas, S., M. Martínez-Pascual, and P.E. Alcaraz. 2013. "Impact of resistance circuit training on neuromuscular, cardiorespiratory and body composition adaptations in the elderly." *Aging and Disease* 4 (5): 256–263.

Rose, G. 1993. *Feminism and Geography: the Limits of Geographical Knowledge*. Cambridge: Polity Press.

Rose, G. 2012. *Visual Methodologies: An Introduction to Researching with Visual Materials, 3rd edition*. Thousand Oaks, London, New Delhi: Sage Publishers.

Rose, N. 1998. *Inventing Our Selves. Psychology, Power and Personhood*. Cambridge: Cambridge University Press.

Rose, N. 2007. *The Politics of Life Itself: Biomedicine, Power, and Subjectivity in the Twenty-First Century*. Princeton and Oxford: Princeton University Press.

Rosenberg, J. 2011. "Whose business is dying? Death, the home and palliative care." *Cultural Studies Review* 17 (1): 15.

Rothman, B.K. 1986. *The Tentative Pregnancy: Prenatal Diagnosis and the Future of Motherhood*. New York: Viking

Roy, P. and J. Connell. 1991. "Climatic change and the future of atoll states." *Journal of Coastal Research* 7 (4): 1057–1075.

Rozanova, J. 2010. "Discourse of successful aging in *The Globe & Mail*: Insights from critical gerontology." *Journal of Aging Studies* 24 (4): 213–222.

Rudman, D.L. 2006. "Shaping the active, autonomous and responsible modern retiree: An analysis of discursive technologies and their links with neo-liberal political rationality." *Ageing and Society* 26 (2): 181–201.

Ryan, J. and H. Montgomery. 2005. "Perspective. Terrorism and the medical response." *The New England Journal of Medicine* 353 (August 11): 543–545.

Said, E. 1993. *Culture and Imperialism*. New York: Vintage Books.

Sargent-Cox, K.A., K.J. Anstey, and M.A. Luszcz. 2012. "The relationship between change in self-perceptions of aging and physical functioning in older adults." *Psychology and Aging* 27 (3): 750–760.

Satariano, W.A., J.M. Guralnik, R.J. Jackson, R.A. Marottoli, E.A. Phelan, and T.R. Prohaska. 2012. "Mobility and aging: New directions for public health action." *American Journal of Public Health* 102 (8): 1508–1515.

Schivelbusch, W. 1986. *The Railway Journey. The Industrialization of Time and Space in the 19th Century.* Oxford: Oxford University Press.

Schlottmann, A. and J. Miggelbrink. 2009. "Visual geographies—an editorial." *Social Geography* 4 (April): 1–11.

Scott, E.S. and L. Steinberg. 2008. "Adolescent development and the regulation of youth crime." *The Future of Children* 18 (2): 15–33.

Scott, P.M. and C. Deane-Drummond. 2006. "Future perfect? Or, what should we hope for?". In C. Deane-Drummond and P.M. Scott (Eds), *Future Perfect? God, Medicine and Human Identity*, 1–12. London and New York: T&T Clark International. A Continuum Imprint.

Scruton, R. 2012. "Timely death." *Philosophical Papers* 41 (3): 421–434.

Scully, J.L., C. Rehmann-Sutter, and R. Porz. 2010. "Human embryos: Donors' and non-donors' perspectives on embryo moral status." In J. Nisker, F. Baylis, I. Karpin, C. McLeod, and R. Mykitiuk (Eds), *The 'Healthy' Embryo. Social, Biomedical, Legal and Philosophical Perspectives*, 16–31. Cambridge: Cambridge University Press.

Shakespeare, W. c.1600. *The Tragedy of Hamlet, Prince of Denmark*, Act III, Scene I. Hosted at The Shakespeare Quartos Archive as Hamlet, 1604. Copy 1. Folger Library, image 8 [http://www.quartos.org/] accessed on 13 December 2013.

Sharpe, S. 2012. "The aesthetics of urban movement: habits, mobility, and resistance." *Geographical Research* 51 (2): 166–172.

Sharples, C. 2006. *Indicative Mapping of Tasmanian Coastal Vulnerabililty to Climate Change and Sea-Level Rise.* Hobart: Consultant Report to the Tasmanian Government Department of Primary Industries and Water Tasmania.

Shaw, J. and M. Hesse. 2010. "Transport, geography and the 'new'mobilities." *Transactions of the Institute of British Geographers* 35 (3): 305–312.

Sheffield, P.E. and P.J. Landrigan. 2011. "Global climate change and children's health: Threats and strategies for prevention." *Environmental Health Perspectives* 119 (3): 291–298.

Sheller, M. 2004. "Automotive emotions: Feeling the car." *Theory, Culture & Society* 21 (4–5): 221–242.

Sheller, M. 2009. "Infrastructures of the imagined island: Software, mobilities, and the architecture of Caribbean paradise." *Environment and Planning A* 41 (6): 1386–1403.

Sheller, M. and J. Urry. 2006. "The new mobilities paradigm." *Environment and Planning A* 38 (2): 207–226.

Sibley, D. 1995. *Geographies of Exclusion. Society and Difference in the West.* London and New York: Routledge.

Sifferlin, A. 2013. "Q&A: World's oldest performing female bodybuilder." *Time Health & Family*, May 30. [http://healthland.time.com/2013/05/30/qa-worlds-oldest-performing-female-bodybuilder/] accessed on 27 December 2013.

Silvey, R. 2004. "Power, difference and mobility: Feminist advances in migration studies." *Progress in Human Geography* 28 (4): 1–17.

Simmel, G. 2005 [1903]. "The metropolis and mental life." In J. Lin and C. Mele (Eds), *The Urban Sociology Reader*, 23–31. London and New York: Routledge.

Simonsen, K. 2003. "The embodied city: From bodily practice to urban life." In J. Öhman and K. Simonsen (Eds), *Voices from the North: New Trends in Nordic Human Geography*, 157–172. Aldershot: Ashgate.

Sixsmith, J., A. Sixsmith, A.M. Fänge, D. Naumann, C. Kucsera, S. Tomsone, M. Haak, S. Dahlin-Ivanoff, and R. Woolrych. 2014. "Healthy ageing and home: The perspectives of very old people in five European countries." *Social Science & Medicine* 106 (2014): 1–9.

Skelton, T. 2010. "Taking young people as political actors seriously: Opening the borders of political geography." *Area* 42 (2): 145–151.

Skirtboarders. 2009. Les Skirtboarders—Boston Dew Tour—Summer 2009 [http://www.youtube.com/watch?v=-qFuEijkSA4] accessed on 14 October 2013.

Soja, E.W. 2009. "Taking space personally." In B. Warf and S. Arias (Eds), *The Spatial Turn: Interdisciplinary Perspectives*, 11–35. London and New York: Routledge.

Stabile, C.A. 1992. "Shooting the mother: Fetal photography and the politics of disappearance." *Camera Obscura* 10 (1 28): 178–205.

Statistics Canada. 2010a. Aboriginal Statistics at a glance—life expectancy. [http://www.statcan.gc.ca/pub/89–645-x/2010001/life-expectancy-esperance-vie-eng.html] accessed on 7 November 2011.

Statistics Canada. 2010b. Life expectancy at birth, by sex, by province (2006–08) [http://www40.statcan.gc.ca/l01/cst01/health26-eng.htm] accessed on 7 November 2011.

Steinberg, P.E. 2005. "Insularity, sovereignty and statehood: The representation of islands on Portolan Charts and the construction of the territorial state." *Geografiska Annaler: Series B, Human Geography* 87 (4): 253–265.

Steinhauser, K.E., N.A. Christakis, E.C. Clipp, M. McNeilly, L. McIntyre, and J.A. Tulsky. 2000. "Factors considered important at the end of life by patients, family, physicians, and other care providers." *The Journal of the American Medical Association* 284 (19): 2476–2482.

Stephens, T. 2010. "What is rhythm in relation to photography?" *Philosophy of Photography* 1 (2): 157–175.

Stevens, L.P., L. Hunter, D. Pendergast, V. Carrington, N. Bahr, C. Kapitzke, and J. Mitchell. 2007. "Reconceptualizing the possible narratives of adolescence." *The Australian Educational Researcher* 34 (2): 107–127.

Stewart, K. 2010. "Afterword. Worlding Refrains." In M. Gregg and G.J. Seigworth (Eds), *The Affect Theory Reader*, 339–353. Durham: Duke University Press.

Stormer, N. 2008. "Looking in wonder: Prenatal sublimity and the commonplace 'life'." *Signs: Journal of Women in Culture and Society* 33 (3): 647–673.

Stratford, E. 1997. "Memory work, geography and environmental studies: Some suggestions for teaching and research." *Australian Geographical Studies* 35 (2): 206–219.

Stratford, E. 1998a. "A biopolitics of population decline: The *Australian Women's Sphere* as a discourse of resistance." *Australian Geographer* 29 (3): 357–370.

Stratford, E. 1998b. "City squares up to youth needs." In *Sunday Tasmanian Hobart, 19 July*. Hobart: Murdoch Press.

Stratford, E. 1998c. "Health and nature in the 19th century Australian women's popular press." *Health & Place* 4 (2): 101–112.

Stratford, E. 2000a. "Exploring gender differences in children's memory of place." In M. Robertson and R. Gerber (Eds), *The Child's World: Triggers for Learning*, 150–170. Melbourne: ACER Press.

Stratford, E. 2000b. "Gender, place and travel: the case of Elsie Birks, South Australian pioneer." *Journal of Australian Studies: Vision Splendid* 66: 116–128

Stratford, E. 2000c. "Skating a fine line. News Focus." *The Mercury,* 30 May.

Stratford, E. 2002. "On the edge: A tale of skaters and urban governance." *Social & Cultural Geography* 3 (2): 193–206.

Stratford, E. 2003. "Editorial. Flows and boundaries: Small island discourses and the challenge of sustainability, community and local environments." *Local Environment* 8 (5): 495–499.

Stratford, E. 2004. "Think global, act local: Scalar challenges to sustainable development of marine environments." In R. White (Ed.), *Controversies in Environmental Sociology*, 150–167. Melbourne: Cambridge University Press.

Stratford, E. 2006a. "Isolation as disability and resource: considering sub-national island status in the constitution of the 'New Tasmania'." *The Round Table: Commonwealth Journal of International Affairs* 95 (386): 575–588.

Stratford, E. 2006b. "Technologies of agency and performance: Tasmania together and the constitution of harmonious island identity." *Geoforum* 37 (2): 273–286.

Stratford, E. 2008. "Islandness and struggles over development: A Tasmanian case study." *Political Geography* 27 (2): 160–175.

Stratford, E. 2009. "Belonging as a resource: the case of Ralphs Bay, Tasmania, and the local politics of place." *Environment and Planning A* 41 (4): 796–810.

Stratford, E., D. Armstrong, and M. Jaskolski. 2003. "Relational spaces and the geopolitics of community participation in two Tasmanian local governments: a case for agonistic pluralism?" *Transactions of the Institute of British Geographers* NS 28 (4): 461–472.

Stratford, E., C. Farbotko, and H. Lazrus. 2013. "Tuvalu, sovereignty and climate change: considering fenua, the archipelago and emigration." *Island Studies Journal* 8 (1): 67–83.

Stratford, E. and A. Harwood. 2001. "Feral travel and the transport field: Some observations on the politics of regulating skating in Tasmania." *Urban Policy and Research* 19 (1): 61–76.

Stratford, E., E. McMahon, C. Farbotko, M. Jackson, and S. Perera. 2011. "Review Forum. Reading Suvendrini Perera's Australia and the Insular Imagination." *Political Geography* 30 (6): 329–338.

Stratford, E. and S. Wells. 2009. "Spatial anxieties and the changing landscape of an Australian airport." *Australian Geographer* 40 (1): 69–84.

Strauss, S. 2012. "Are cultures endangered by climate change? Yes, but . . ." *Wiley Interdisciplinary Reviews: Climate Change* 3 (4): 371–377.

Stutzer, A. and B.S. Frey. 2008. "Stress that doesn't pay: The commuting paradox." *Scandinavian Journal of Economics* 110 (2): 339–366.

Sullivan, S.E. 2010. "Management's unsung theorist: An examination of the works of Lillian M. Gilbreth." *Biography* 18 (1): 31–41.

Sutton-Smith, B. 2009. *The Ambiguity of Play*. Boston: Harvard University Press.

Svendsen, M.N. and L. Koch. 2008. "Unpacking the 'spare embryo'. Facilitating stem cell research in a moral landscape." *Social Studies of Science* 38 (1): 93–110.

Tasmanian Parliament. 2001. Traffic Amendment (Wheeled Recreational Devices and Wheeled Toys) Bill 2011 No. 17. Second Reading Thursday 29 March. [http://www.parliament.tas.gov.au/ParliamentSearch/isysquery/59856bd2-d533–4ccb-baa1-e0a953ec8948/4/doc/] accessed on 16 October 2013.

Tasmanian Police. 2013. Serious skateboard crash Hobart. [http://www.police.tas.gov.au/news/posts/view/4110/serious-skateboard-crash-hobart/] accessed on 15 October 2013.

Taylor, F.W. 1911. *The Principles of Scientific Management*. New York: Harper.

Temelová, J. and J. Novák. 2011. "Daily street life in the inner city of Prague under transformation: The visual experience of socio-spatial differentiation and temporal rhythms." *Visual Studies* 26 (3): 213–228.

The Advance Care Directive Association Inc. 2014. What is advance care planning? [http://www.advancecaredirectives.org.au/] accessed on 26 March 2014.

The British Museum. 2014. Explore / Online tours: Cradle to Grave by Pharmacopoeia. [http://www.britishmuseum.org/explore/online_tours/museum_and_exhibition/audio_description_tour/cradle_to_grave_by_pharmacopoe.aspx] accessed on 10 February 2014.

The Healing Foundation UK. 2004–2011. Ambassadors. [http://www.theheal ingfoundation.org.uk/thf2008/ambassadors.htm] accessed on 19 March 2014.

The Matthew Fulham Foundation. 2012. Fundraising. [http://www.matthewfulham. org/Fundraising.html] accessed on 20 March 2014.

The Order of Australia Association. 2014. Order of Australia. [http://www. theorderofaustralia.asn.au/] accessed on 21 March 2014.

The President's Council on Bioethics. 2002. Human Cloning and Human Dignity: An Ethical Inquiry. [http://bioethics.georgetown.edu/pcbe/reports/cloningreport/ research.html] accessed on 24 January 2013.

The University of Georgia (2012). Overview. The Geoge Foster Peabody Awards. [http://peabodyawards.com/about-the-peabody/overview-history/] accessed on 2 March 2013.

Thompson, C. 2010. "Informed consent for the age of pluripotency and embryo triage: From alienation, anonymity and altruism to connection, contact and care." In J. Nisker, F. Baylis, I. Karpin, C. McLeod and R. Mykitiuk (Eds), *The 'Healthy' Embryo. Social, Biomedical, Legal and Philosophical Perspectives*, 45–60. Cambridge: Cambridge University Press.

Thomson, F. 2007. "Are methodologies for children keeping them in their place?" *Children's Geographies* 5 (3): 207–218.

Tillett, T. 2011. "Climate change and children's health: Protecting and preparing our youngest." *Environmental Health Perspectives* 119 (3): 132.

Tompkins, E.L. 2005. "Planning for climate change in small islands: Insights from national hurricane preparedness in the Cayman Islands." *Global Environmental Change* 15 (2): 139–149.

Traxler, M. 2002. "Fair chore division for climate change." *Social Theory and Practice* 28 (1): 101–134.

Trimbur, L. 2011. " 'Tough love': Mediation and articulation in the urban boxing gym." *Ethnography* 12 (3): 334–355.

Tronto, J. 1987. "Beyond Gender Difference to a Theory of Care." *Signs: Journal of Women in Culture and Society* 12 (4): 644–663.

Tronto, J. 1993. *Moral Boundaries: A Political Argument for an Ethic of Care*. New York: Routledge.

Tuan, Y.-F. 1999. "Geography and evil: A sketch." In J.D. Proctor and D.M. Smith (Eds), *Geography and Ethics: Journeys in a Moral Terrain*, 106–119. London and New York: Routledge.

Tuan, Y.-F. 2001. *Space and Place: The Perspective of Experience*. Minneapolis: University of Minnesota Press.

Tucker, F. 2003. "Sameness or difference? Exploring girls' use of recreational spaces." *Children's Geographies* 1 (1): 111–124.

UK Government. Department of Transport. 2011. Transport Statistics Great Britain 2011. [https://www.gov.uk/government/uploads/system/uploads/attachment_ data/file/11833/Transport_Statistics_Great_Britain_2011___all_chapters.pdf] accessed on 18 March 2014.

UK Office for National Statistics. 2006. *National Statistics. Focus on Health*. London: UK Office for National Statistics.

Ullman, E.L. 1953. "Human geography and area research." *Annals of the Association of American Geographers* 43 (1): 54–66.

Umstattd, M.R. and J. Hallam. 2007. "Older adults' exercise behavior: Roles of selected constructs of social-cognitive theory." *Journal of Aging and Physical Activity* 15 (2): 206–218.

United Nations Children's Fund (UNICEF). 2010. *Core Commitments for Children in Humanitarian Action*. New York: Humanitarian Policy Section, Office of Emergency Programs and Division of Communication, UNICEF.

UNICEF, World Health Organization, UN Population Division and The World Bank. 2004–2014. Child Mortality Estimates Information. [http://www.childmortality.org/index.php?r=site/index&language=] accessed on 16 March 2014.

United Kingdom. Security Service MI5. nd. Terrorism. [https://www.mi5.gov.uk/home/the-threats/terrorism.html] accessed on 27 November 2013.

United Methodist TV. 2012. World Record Female Bodybuilder Ernestine Shepherd, YouTube. [http://www.youtube.com/watch?v=elsOwJ4IyyA] accessed on 6 January 2014.

United Nations. 1992. *United Nations Framework Convention on Climate Change.* New York: United Nations.

United Nations. Department of Economic and Social Affairs. Population Division. 2002. *World Population Ageing: 1950–2050.* New York: UNDESA.

United Nations. Department of Economic and Social Affairs. Population Division. 2013. *World Population Ageing 2013.* New York: UNDESA.

United Nations. Office of the High Commissioner for Human Rights. 1989. Convention on the Rights of the Child. Adopted and opened for signature, ratification and accession by General Assembly resolution 44/25 of 20 November 1989 with entry into force 2 September 1990, in accordance with article 49. [http://www.ohchr.org/EN/ProfessionalInterest/Pages/CRC.aspx] accessed on 16 July 2013.

United Nations, High Commission for Refugees. 1951. *Text of the 1951 Convention Relating to the Status of Refugees. Text of the 1967 Protocol Relating to the Status of Refugees. Resolution 2198 (XXI) adopted by the United Nations General Assembly with an Introductory Note by the Office of the United Nations High Commissioner for Refugees.* Geneva: UNHCR.

United States Government. 2001. "Uniting and Strengthening American by Providing Appropriate Tools Required to Intercept and Obstruct Terrorism (USA PATRIOT) Act. 115 Stat. 272. Public Law 107–56-Oct. 26 2001.

United States Government. nd. US Code, Chapter 113B: Terrorism. [http://codes.lp.findlaw.com/uscode/18/I/113B/2331] accessed on 27 November 2013.

United States Government. Ratified 1788. The United States Constitution. [http://constitutionus.com/] accessed on 27 November 2013.

United States Government. Department of Health and Human Services. Centers for Disease Control and Prevention and National Center for Health Statistics. 2013. Injuries and violence are leading causes of death: key data and statistics. [http://www.cdc.gov/injury/overview/data.html] accessed on 16 March 2014.

United States Government. Department of Health and Human Services. Centers for Disease Control and Prevention and National Center for Health Statistics. 2010. Health, United States, 2010, With Special Feature on Death and Dying. [http://www.ncbi.nlm.nih.gov/books/NBK54374/#specialfeature.s8] accessed on 23 March 2014.

United States Government. Department of Transportation. 2013. Research and Innovative Technology Administration. Bureau of Transport Statistics. National Transportation Statistics. Table 2–4: Distribution of Transportation Fatalities by Mode. [http://www.rita.dot.gov/bts/sites/rita.dot.gov.bts/files/publications/national_transportation_statistics/html/table_02_04.html] accessed on 17 March 2014.

United States Government. Internal Revenue Service. 2012. Publication 463: 4. Transportation. Car expenses. [http://www.irs.gov/publications/p463/ch04.html] accessed on 13 November 2013.

University of Essex. Institute for Social and Economic Research. 2013. British Household Panel Survey. [https://www.iser.essex.ac.uk/bhps] accessed on 16 November 2013.

Urbano, M., N. Maclellan, T. Ruff, and G. Blashki. 2010. *Climate Change and Children in the Pacific Islands. Report Submitted to UNICEF Pacific.* Melbourne: University of Melbourne Nossal Institute for Global Health.

Urry, J. 1999. Automobility, Car Culture and Weightless Travel: A discussion paper. [http://www.comp.lancs.ac.uk/sociology/soc008ju.html] accessed on 6 August 2002.

Urry, J. 2000. *Sociology Beyond Societies: Mobilities for the Twenty-First Century.* London and New York: Routledge.

Urry, J. 2006. "Inhabiting the car." *The Sociological Review* 54 (Supplement S1): 17–31.

Urry, J. 2007. *Mobilities.* Cambridge: Polity Press.

Valentine, G. 1996. "Angels and devils: Moral landscapes of childhood." *Environment and Planning D: Society and Space* 14 (5): 581–599.

Valentine, G., T. Skelton, and D. Chambers. 1998. "Cool places: An introduction to youth and youth cultures." In T. Skelton and G. Valentine (Eds), *Cool Places. Geographies of Youth Cultures,* 1–32. London and New York: Routledge.

Van Geenhuizen, M. and P. Nijkamp. 2003. "Coping with uncertainty: An expedition into the field of new transport technology." *Transportation Planning and Technology* 26 (6): 449–467.

Vannini, P. 2011. "Mind the gap: The *Tempo Rubato* of dwelling in lineups." *Mobilities* 6 (2): 273–299.

Vannini, P., G. Baldacchino, L. Guay, S.A. Royle, and P.E. Steinberg. 2009. "Recontinentalizing Canada: Arctic ice's liquid modernity and the imagining of a Canadian archipelago." *Island Studies Journal* 4 (2): 121–138.

Veitayaki, J. 2010. "Climate change adaptation issues in small island developing states." In J. Pulhin and J. Pereira (Eds), *Climate Change Adaptation and Disaster Risk Reduction: Issues and Challenges,* 369–391. Bradford: Emerald Group Publishing.

Vertesi, J. 2008. "Mind the gap: The London Underground Map and users' representations of urban space." *Social Studies of Science* 38 (1): 7–33.

Viry, G., V. Kaufmann, and E.D. Widmer. 2009. "Social integration faced with commuting: More widespread and less dense support networks." In T. Ohnmacht, H. Maksim, and M.M. Bergman (Eds), *Mobilities and Inequality,* 121–143. Aldershot: Ashgate Publishing.

Vivoni, F. 2009. "Spots of spatial desire: Skateparks, skateplazas, and urban politics." *Journal of Sport and Social Issues* 33 (2): 130–149.

Waitt, G. and L. Cook. 2007. "Leaving nothing but ripples on the water: Performing ecotourism natures." *Social & Cultural Geography* 8 (4): 535–550.

Waitt, G. and T. Harada. 2012. "Driving, cities and changing climates." *Urban Studies* 49 (15): 3307–3325.

Waitt, G., E. Ryan, and C. Farbotko. 2014. "A visceral politics of sound." *Antipode* 46 (1): 283–300.

Walker, G. 2012. *Environmental Justice: Concepts, Evidence and Politics.* Abingdon and New York: Routledge.

Walker, J., P. Orpin, H. Baynes, E. Stratford, K. Boyer, N. Mahjouri, C. Patterson, A. Robinson, and J. Carty. 2012. "Insights and principles for supporting social engagement in rural older people." *Ageing and Society* 33 (6): 938–963.

Wallis, L. 2013. "Is 25 the new cut-off point for adulthood?" *BBC News Magazine,* 23 September. [http://www.bbc.co.uk/news/magazine-24173194] accessed on 23 September 2013.

Waters, T. 2013. Skateboarding fatality report for the USA, 2012. Skaters For Public Skateparks. [http://www.skatepark.org/park-development/2013/03/2012-lskate boarding-fatalities/] accessed on 2 February 2014.

Watson, J. and F. Crick. 1953. "Molecular structure of nucleic acids." *Nature* 171 (4356): 737–738.

Weightman, G., S. Humphries, J. Mack, and J. Taylor. 2007. *The Making of Modern London.* London: Ebury Press [Random House].

Weller, S. 2006. "Situating (young) teenagers in geographies of children and youth." *Children's Geographies* 4 (1): 97–108.

Welsh, D. 2010. *Underground Writing: The London Tube from George Gissing to Virginia Woolf.* Liverpool: Liverpool University Press.

Wener, R.E., G.W. Evans, D. Phillips, and N. Nadler. 2003. "Running for the 7:45: The effects of public transit improvements on commuter stress." *Transportation* 30 (2): 203–220.

Westcott, W. and T.R. Baechle. 2007. *Strength Training Past 50.* Champaign, IL: Human Kinetics.

White, A. and R. Bluhm. 2010. "Embryo health and embryo research." In J. Nisker, F. Baylis, I. Karpin, C. McLeod, and R. Mykitiuk (Eds), *The 'Healthy' Embryo. Social, Biomedical, Legal and Philosophical Perspectives,* 187–199. Cambridge: Cambridge University Press.

Wolf, S.A. 2006. "The Mermaid's Purse: Looking closely at young children's art and poetry." *Language Arts* 84 (1): 10–20.

Wolmar, C. 2004. *The Subterranean Railway: How the London Underground was Built and How it Changed the City Forever.* London: Atlantic Books.

Woodyer, T. 2012. "Ludic geographies: Not merely child's play." *Geography Compass* 6 (6): 313–326.

Woolley, H., T. Hazelwood, and I. Simkins. 2011. "Don't skate here: Exclusion of skateboarders from urban civic spaces in three northern cities in England." *Journal of Urban Design* 16 (4): 471–487.

World Health Organization. 2012. Global health observatory: Infant mortality. Situation and trends. [http://www.who.int/gho/child_health/mortality/neonatal_infant/en/] accessed on 16 March 2014.

World Health Organization. 2013. The top 10 causes of death, Fact Sheet No.310. [http://www.who.int/mediacentre/factsheets/fs310/en/] accessed on 17 March 2014.

World Health Organization. 2014a. Children: reducing mortality. Fact Sheet No. 178, Updated September 2013. [http://www.who.int/mediacentre/factsheets/fs178/en/] accessed on 16 March 2014.

World Health Organization. 2014b. Global health observatory: life expectancy. [http://www.who.int/gho/mortality_burden_disease/life_tables/situation_trends_text/en/] accessed on 22 March 2014.

Wright, C. 2013. "Against flourishing: Wellbeing as biopolitics, and the psychoanalytic alternative." *Health, Culture and Society* 5 (1): 20–35.

Wright, E.O. 2010. *Envisioning Real Utopias.* London: Verso.

Wright, E.O. 2013. "Transforming capitalism through real utopias." *American Sociological Review* 78 (1): 1–25.

Wright, F.L. 1945. *When Democracy Builds.* Chicago: University of Chicago Press.

Yamamoto, L. and M. Esteban. 2010. "Vanishing island states and sovereignty." *Ocean & Coastal Management* 53 (1): 1–9.

Yeung, J., S. Wearing, and A.P. Hills. 2008. "Child transport practices and perceived barriers in active commuting to school." *Transportation Research Part A: Policy and Practice* 42 (6): 895–900.

Zelinsky, W. 1994. "Gathering places for America's dead: How many, where, and why?" *The Professional Geographer* 46 (1): 29–38.

Index

9780415659369

An environmentally friendly book printed and bound in England by www.printondemand-worldwide.com

PEFC Certified

This product is
from sustainably
managed forests
and controlled
sources

www.pefc.org

PEFC/16-33-415

This book is made of chain-of-custody materials; FSC materials for the cover and PEFC materials for the text pages.

#0027 - 171115 - C0 - 229/152/12 [14] - CB - 9780415659369